Evolution of Dental Tissues and Paleobiology in Selachians

Vertebrate Paleobiology and Paleoenvironments Set

coordinated by
Eric Buffetaut

Evolution of Dental Tissues and Paleobiology in Selachians

Gilles Cuny
Guillaume Guinot
Sébastien Enault

First published 2017 in Great Britain and the United States by ISTE Press Ltd and Elsevier Ltd

ISTE Press Ltd
27-37 St George's Road
London SW19 4EU
UK

www.iste.co.uk

Elsevier Ltd
The Boulevard, Langford Lane
Kidlington, Oxford, OX5 1GB
UK

www.elsevier.com

Notices

Knowledge and best practice in this field are constantly changing. As new research and experience broaden our understanding, changes in research methods, professional practices, or medical treatment may become necessary.

Practitioners and researchers must always rely on their own experience and knowledge in evaluating and using any information, methods, compounds, or experiments described herein. In using such information or methods they should be mindful of their own safety and the safety of others, including parties for whom they have a professional responsibility.

To the fullest extent of the law, neither the Publisher nor the authors, contributors, or editors, assume any liability for any injury and/or damage to persons or property as a matter of products liability, negligence or otherwise, or from any use or operation of any methods, products, instructions, or ideas contained in the material herein.

For information on all our publications visit our website at http://store.elsevier.com/

British Library Cataloguing-in-Publication Data
A CIP record for this book is available from the British Library
Library of Congress Cataloging in Publication Data
A catalog record for this book is available from the Library of Congress
ISBN 978-1-78548-139-0

Printed and bound in the UK and US

Contents

Introduction vii

Chapter 1. Mineralized Tissues 1

   1.1. Dental structure and tissues 1
   1.2. Enameloid: a problematic tissue 2
   1.3. Histology and variation 3
   1.4. Developmental origin
   and mineralization 7
   1.5. Dentin. 14
   1.6. Homology of enameloid tissues 16

Chapter 2. Paleozoic Sharks 19

   2.1. Do all Paleozoic shark teeth
   possess enameloid? 19
   2.2. Ctenacanth sharks 28

Chapter 3. Hybodont Sharks 33

   3.1. Enameloid microstructure in
   hybodont teeth 34
   3.2. Development of specialized tooth
   structures and associated changes in
   the enameloid microstructure 39
   3.3. The mysterious *Ptychodus* 43
   3.4. Conclusion. 45

Chapter 4. Enameloid Microstructure in Rays   47

4.1. The phylogenetic position of skates
and rays within the neoselachians.   47
4.2. Historical context   48
4.3. Batomorph enameloid: diversity
and evolution.   49
4.4. Durophagy and other
trophic specializations.   54

Chapter 5. Enameloid Microstructure
Diversity in Modern Shark Teeth   61

5.1. Diversity and evolution of structures   62
5.2. Serrated teeth and mega-predation   73
5.3. Example of adaptation to durophagy:
bullhead sharks   77

Chapter 6. Comparison of Enameloid Microstructure
in Actinopterygian and Elasmobranch Teeth   81

6.1. Ganoine and acrodine   81
6.2. Comparison of acrodin and
elasmobranch enameloid   84

Conclusion   97

Glossary   105

Bibliography   111

Index   127

Introduction

The appearance of jaws represents a major event in the history of vertebrates and marks the transition between the agnathans (jawless "fish") and the gnathostomes (animals possessing a true articulated jaw). This new body part, combined with dental structures, gave access to a whole new range of ecological niches for animals which had previously been limited mainly to filter feeding. With the appearance of jaws, alimentary systems based on the ingestion of large food items or prey protected by a carapace or shell became possible, resulting in significant diversification of the gnathostomes during the Silurian Period between 443 and 419 million years ago (King *et al.*, 2016).

In addition to articulated jaws, gnathostomes are characterized by a range of dental structures that varies widely in terms of morphology and function. The association between teeth and jaws appears natural; without teeth, what is the point of possessing jaws? Teeth lie, in effect, at the interface between the latter and food. These teeth are an important source of information, both in order to understand physiological processes in organisms and evolutionary changes on the geologic time scale. While the first level of study of a tooth is morphological, insofar as these characteristics can often be seen by the naked eye, other characteristics are likely to be just as informative, particularly the tooth chemical composition or the structure of the tissues that constitute it, which enable the tooth to perform its function in an optimal manner. To manipulate food effectively, the tooth must be resistant, so it is covered in hypermineralized tissue, i.e. a tissue that is at least 96% mineral in composition with no more than 4% of organic matter. This tissue is referred to as enamel in humans and in the vast majority of tetrapods (animals with limbs). It is different somewhat in fish, where this tissue is more often designated by the general term enameloid. It is important to remember here that the term fish in biology does not designate a natural group, called a clade, which includes all the descendants of a single common ancestor, but rather a disparate group of very different animals with very little relationship to one another. A trout is in fact more similar to an elephant, with which it shares a bony skeleton, than to a shark.

The microcrystals that form enamel or enameloid can be arranged in different ways, forming what we call microstructures. In modern selachians and fossils (the term *selachian* is preferable to *shark* as it includes skates and rays), these microstructures show surprising diversity and help us to better understand the phylogenetic affinities between fossil teeth found in isolation and the adaptation to certain feeding systems of their owners. We can thus reconstruct the lifestyles of animals that have been extinct for hundreds of millions of years. These characteristics are well known in mammals and have been the subject of study for many years. In this book, we will review these different microstructures in selachians and examine what they can tell us about the evolution and the past biology of these animals, as well as what they reveal to us about the evolution of teeth themselves.

These questions are of particular interest insofar as the skeletons of modern and fossil Chondrichthyes are composed mainly of cartilage that is more or less calcified depending on structure and species, a tissue that does not fossilize well. Though complete fossils have been found in several fossiliferous beds, they remain exceptionally rare, and the vast majority of the fossil record of Chondrichthyes is made up of isolated teeth and dermal denticles (unlike bony vertebrates, in which the information provided by dental features may be complemented by the rest of the skeleton, which is preserved much more often in the fossil record). In light of this fact, it is not hard to understand the importance of the morphology and structure of chondrichthyan teeth to our understanding of their relationships and evolutionary history.

1

Mineralized Tissues

1.1. Dental structure and tissues

The typical structure of a chondrichthyan tooth includes a base (or root), allowing it to be fixed to the cartilage of the jaw, and a crown composed of dentin and covered with enameloid (Figure 1.1). The variability in the tissue composition of these various structures was recognized very early and has generated a considerable amount of study with occasionally contradictory results. Histological characteristics have been and are still used in combination with the morphological characteristics of dentition, enabling in some cases the identification and classification of fossil taxa with unclear affinities. Numerous studies followed on the heels of those of Agassiz (1833) and Owen (1840), aimed at clarifying the systematic and functional interests of histological and microstructural characteristics. These questions are still not wholly resolved, due largely to a sampling that remains too small to identify the range of variation of histological characteristics satisfactorily and to a poor understanding of the developmental mechanisms of the origin of these tissues (Enault *et al.*, 2015).

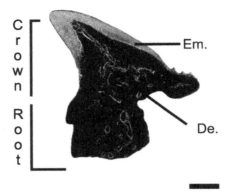

Figure 1.1. *Cross-section of* Heterodontus *sp., (between sp. and anterior)* anterior tooth. *The scale bar is 0.5 cm. Em.: Enameloid, De.: Dentin*

1.2. Enameloid: a problematic tissue

Most of the controversies surrounding the dental tissues of Chondrichthyes have to do with the hypermineralized tissue (i.e. tissue that is ≤5% organic in composition) covering the crown of the teeth, the enameloid. Its nature and developmental origin have been debated for more than a century (Prostak and Skobe, 1988). Given its topology and degree of mineralization, it has been considered by some as homologous to enamel (Moss *et al.*, 1964); however, there are fundamental differences between these two types of tissue. Enamel is produced by an epithelium, a tissue composed of polarized, closely connected cells, through the mineralization of an extracellular matrix that completely lacks collagen and is composed of specific proteins. These proteins, called EMP (*extracellular matrix proteins*), are secreted by ameloblasts, which differentiate themselves from the dental epithelium (Kawasaki, 2009a; Sire *et al.*, 2005, 2007). Its incremental formation is posterior to that of the underlying dentin, the protein matrix of which is produced by odontoblasts, mesenchymal cells that do not have the same embryonic origin as the epithelium (Sharpe, 2001). Enameloid, on the contrary, has often been described as a tissue of mixed developmental origin, produced by both ameloblasts and odontoblasts, although the respective contributions of each of these types of cells remain poorly understood. Its extracellular matrix is described as being rich in collagen, which, along with the often ill-defined boundary between the two tissues, has led some authors to consider it as a particularly mineralized form of dentin. Unlike enamel, its mineralization is anterior to that of the dentin, another difference indicating a probable convergence between the two tissues (Kemp, 1999; Tomes, 1898).

In Chondrichthyes, the main component of enameloid is fluorapatite, a calcium phosphate rich in fluorine, which forms microcrystals (Chen *et al.*, 2014; Francillon-Vieillot *et al.*, 1990). This micrometric crystalline structure, or microstructure, has generated a significant interest, particularly on the part of paleontologists and systematicians, because of its seeming taxonomic advantages (Cappetta, 2012; Cuny and Benton, 1999; Reif, 1977, 1978, 1980b; Zangerl, 1981). The same is true for the arrangement of hydroxyapatite microcrystals forming enamel in mammals that is currently used by paleontologists. The microstructural characteristics of enameloid are also widely used for Chondrichthyes, although they are only described in a very limited number of taxa relative to the substantial diversity of the group (Enault *et al.*, 2015).

1.3. Histology and variation

Histological variations in the dental microstructure of Chondrichthyes have long been considered a tool enabling the differentiation of the teeth of neoselachians (formed of batomorphs, rays, skates and torpedoes, and selachimorphs, including the so-called modern sharks; see Figure 1.2) from those of all other chondrichthyans in the fossil record. The work of the German paleontologist Wolf-Ernst Reif in the

1970s showed the existence of a triple layer of enameloid proper to neoselachians (Figure 1.3). This is characterized by (i) an external layer, the SLE (*shiny layer enameloid*), composed of individualized, randomly oriented fluorapatite crystals thought to limit the formation and propagation of microfissures on the surface of the tooth. This layer, which is particularly thin and sensitive to acid etchings, is usually difficult to see. (ii) A middle layer, the PBE (*parallel bundled enameloid*), which offers resistance to tensile forces and of which the microcrystals are arranged in more or less converging lines parallel to the surface of the crown. (iii) An internal layer, the TBE (*tangled bundled enameloid*), whose lines of crystals form less well-defined bundles lacking apparent orientation. The TBE is thought to offer resistance to compressive forces, which is perhaps illustrated by its importance in the teeth of certain durophagous taxons such as *Heterodontus* (Gillis and Donoghue, 2007; Thies, 1982). It is important to note, however, that these functional interpretations remain entirely theoretical and have yet to be tested in a strict experimental setting.

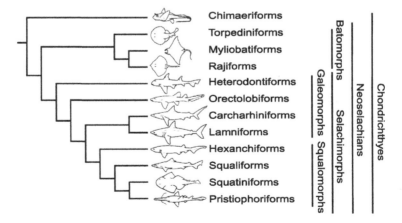

Figure 1.2. *Cladogram illustrating the hypothetical relationships between the orders of modern Chondrichthyes, modified after Douady et al. (2003); McEachran and Aschliman (2004). Other topologies exist as well (e.g. Guinot et al., 2012; Maisey et al., 2004)*

In any case, this characteristic structure was then considered to be absent from the enameloid of other elasmobranchs, particularly those in the neoselachians' sister group, the hybodonts. The teeth of the latter were supposed to have a non-differentiated microstructure composed of isolated microcrystals devoid of a preferential orientation. This single enameloid layer, called SCE (*single-crystallite enameloid*), was then considered homologous to the most external layer of neoselachian enameloid, the SLE (Andreev and Cuny, 2012). Curiously, the characteristic "triple layer" structure also seemed to be absent from the enameloid of rays, which are members of the neoselachian group. This particularity has long been attributed to their diet (Preschoft *et al.*, 1974), and will be addressed in the chapter of this book dedicated to rays.

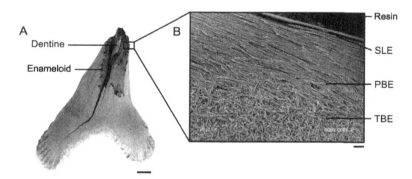

Figure 1.3. *"Triple layer" microstructure of neoselachian enameloid. A: nursehound tooth (*Scyliorhinus stellaris*). The scale bar is 0.5 mm. B: Cross-section of the tooth of* Odontaspis *sp.† under SEM to show the orientation of the bundles of the PBE. The scale bar is 10 µm*

A recent re-examination of the diversity of dental microstructures existing in a wide sampling of neoselachians has shown the limits of the terminology used, demonstrating that it is no longer adequate to describe all of the morphological variation in enameloid encountered in this group. Thus, instead of a system based on the variation in the number of differentiated layers within the enameloid (from one for non-neoselachian taxons and rays to three for selachimorphs), a slightly more complex system has been proposed (Enault *et al.* 2015) by the authors of this book (Figure 1.4), one based on two main units: the SCE, which combines the layer of the same name and the SLE described above (still characterized by a lack of differentiation and organization), and the BCE (*bundled crystallite enameloid*). This group includes three distinct units, the PBE and TBE described above and the RBE (*radial bundled enameloid*), used to describe the radial bundles of microcrystals (oriented perpendicularly to the surface of the tooth) that are more or less well defined and present morphologies which appear to be highly variable and are present in most of the groups discussed here (selachimorphs, batomorphs and hybodonts, but are apparently absent in ctenacanths). Although the combination of all the elements of this structure seems to be found only in selachimorphs, it nevertheless makes it possible to describe all the dental microstructure variants found in elasmobranchs.

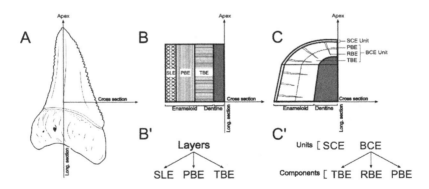

Figure 1.4. *Evolution of the understanding of the dental microstructure of neoselachians. A: Tooth of* Carcharodon carcharias *illustrating the section views currently used. B-B': Schematic conception of the structure of the enameloid of neoselachians following the work of Reif (1973). C-C': Current conception based on a wider sampling of taxons, particularly batomorphs. Modified from Enault et al. (2013; 2015)*

1.4. Developmental origin and mineralization

Besides the problem of the still too limited sampling of the chondrichthyans as a whole, which makes it difficult to target histological variations in enameloid, the gray areas relative to its developmental origins also limit our understanding of its evolution. Most of the controversies surrounding enameloid involve two main points, its cellular origin and the nature of its protein matrix, questions that have been the subject of debate for over a century (Cappetta, 2012; Peyer, 1968; Prostak and Skobe, 1988).

Figure 1.5. *Successive stages of dental development in vertebrates. Modified from Debiais-Thibaud et al., 2011*

The use of the term *enameloid* (Orvig, 1967; Poole, 1967, 1971) is problematic in itself, insofar as it was initially introduced in order to differentiate hypermineralized tissues in mammals and "fish" in the general sense of the term. Consequently, although enamel designates a very specific type of tissue, the term enameloid is often used to describe a variety of hypermineralized tissues existing in non-tetrapod vertebrates (but also in some amphibians), whose homology relationships are considered dubious by many authors (e.g. Zangerl *et al.*, 1993).

In answering this question, it is useful to provide a quick recapitulation of dental development in vertebrates (Figure 1.5). Odontogenesis begins with the formation of a placode, which involves a thickening of the epithelium lying on top of mesenchymal cells (Figure 1.5(a)). This epithelium then folds in on itself to form a tooth bud (Figure 1.5(b)), the base of which becomes stronger in the final stages (Figure 1.5(c)). Genetic mechanisms then cause the differentiation of the cells of the epithelium into amelobasts and the cells of the mesenchyme into odontoblasts. Once these cells have differentiated, they secrete an extracellular matrix at the interface of the two types of cells, and it is the mineralization of this matrix that makes it possible to produce functional tissues (Figure 1.5(d)). This is a process shared by most vertebrates (e.g. Debiais-Thibaud *et al.*, 2011).

In this chapter, we will try to contribute new elements to these questions, based on the existing literature and on some recent data generated by developmental biology. The involvement of odontoblasts in the synthesis of the enameloid protein matrix has been recognized for many years (Grady, 1970; Kvam, 1950; Peyer, 1968; Tomes, 1898). As mentioned at the beginning of this chapter, this observation has spurred several authors to consider enameloid as a particularly hardened form of dentin (Kvam, 1946; 1950; Lübke *et al.*, 2015; Mader, 1986; Schmidt, 1958; Owen, 1840; Zangerl *et al.*, 1993). These authors have come up with their own terminologies for this substance (e.g. "durodentin", "vitrodentin" and "mesodermic enamel") until they have become virtually inextricable, lacking a unifying and concerted approach to the nature of this tissue. To illustrate these disagreements, we will cite several hypotheses given as to the nature of enameloid, describing it as a tissue in which:

– the matrix is characterized by a non-collagenic protein secreted by odontoblasts, the organic framework of which is mineralized by ameloblasts (Tomes, 1898);

– the matrix is collagenic in nature, secreted by odontoblasts and mineralized by ameloblasts (Kerebel *et al.* 1970; Grady, 1970);

– the matrix is partly composed of collagen, which interferes with other proteins secreted by ameloblasts, before the mineralization of the organic framework (Poole, 1971; Shellis, 1978);

– the matrix is composed of tubules of enameline (an EMP protein found in the enamel of Osteichthyes), secreted by ameloblasts and forming "giant fibers" (Kemp, 1985);

– a variant of this hypothesis proposes that if the enameline tubules were epithelial in origin, the "giant fibers" would actually be produced by odontoblasts (Goto, 1978; Prostak and Skobe, 1988).

Most of these hypotheses acknowledge the activity of odontoblasts, which would be responsible for the secretion of the collagenic framework of the enameloid matrix. This aspect of the development of enameloid is no longer called into question by modern researchers, having been demonstrated by multiple studies on the basis of numerous imaging and histochemical techniques, for example.

The precise role of ameloblasts, though, is less clear. Significantly, their potential involvement in the synthesis of the enameloid matrix and their contribution to the mineralization of tissue remains to be clarified. In taxa possessing enamel, ameloblasts participate in both the synthesis of its matrix by synthesizing specific proteins (the EMP already addressed) and in the maturation of tissues by synthesizing other proteins causing its mineralization while breaking down the organic framework. Although some EMP have been identified in the enameloid of several chondrichthyans, these results are mainly based on immunohistological techniques using non-specific antibodies (Herold, Graver and Christner, 1980; Satchell *et al.*, 2002), a practice that raises the possibility of false-positives, particularly as other authors, basing their work on the same approach (Ishiyama *et al.*, 1994), have shed no further light on the involvement of ameloblasts, which is a warning to exercise caution.

However, it has recently been shown in several non-chondrichthyan taxa possessing enameloid (teleosteans, amphibians) that ameloblasts are also capable of synthesizing collagens, thus participating actively in the formation of the enameloid matrix in these species (Assaraf-Weill *et al.*, 2014; Huysseune *et al.*, 2008; Kawasaki *et al.*, 2005; Kawasaki, 2009). These collagens are particularly important because they constitute the most important structural proteins in the extracellular matrices of vertebrates, not only enabling the nucleation of hydroxyapatite crystals during the mineralization process, but also affecting their orientation within the matrix (Hall, 2005; Landis and Silver, 2009).

Figures 1.6 and 1.7 illustrate the histology of developing dentition in *Scyliorhinus canicula* (a selachimorph) and *Raja clavata* (a batomorph). Four distinct colorations are used: hematoxylin eosin saffron (HES), masson's trichrome (TM), reticulin stain (RET) and periodic acid Schiff-alcian blue (PAS-AB). These markings help identify the main components of the extracellular matrices, particularly the collagens.

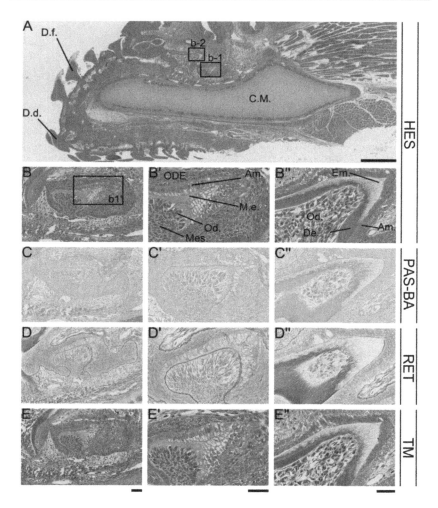

Figure 1.6. *Dental development of S. canicula (juvenile, 20 cm). A: Longitudinal section of lower jaw illustrating dermal denticles, functional teeth and several tooth buds. HES marking. The scale bar is 500 μm. B–E: tooth bud 1 (b-1). B'–E': Enlargement of the cuspid of the first bud. B''–E'': Enlargement of the cuspid of the second bud (b-2). The scale bar is 50 μm. Am: Ameloblasts; C.M.: Meckel's Cartilage; De: Dentin; D.d.: Dermal denticle; D.f.: Functional tooth; Em: Enameloid; M.e.: Enameloid matrix; Mes: Mesenchyme; Od: Odontoblasts. For a color version of this figure, see www.iste.co.uk/cuny/selachians.zip*

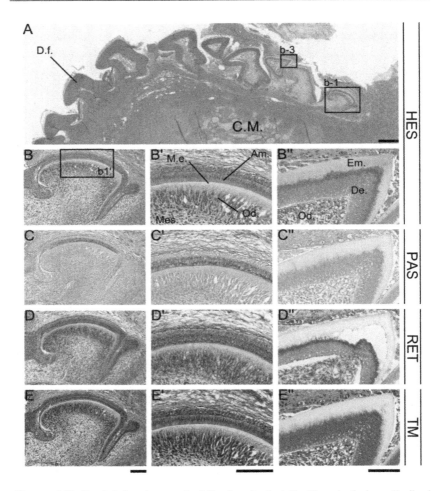

Figure 1.7. *Dental development of* R. clavata *(adult, 45 cm). A: Longitudinal section of the lower jaw showing functional teeth and several tooth buds. HES marking. The scale bar is 500 µm. B–E: Tooth bud 1 (b-1). B'–E': Enlargement of the ameloblasts/odontoblasts junction of the first bud. B"–E": Enlargement of cuspid of the third bud (b-2). The scale bar is 100 µm. Am: Ameloblasts; C.M.: Meckel's Cartilage; De: Dentin; D.f.: Functional tooth; Em: Enameloid; M.e.: Enameloid matrix; Mes: Mesenchyme; Od: Odontoblasts. For a color version of this figure, see www.iste.co.uk/cuny/selachians.zip*

Histologically speaking, dental development is remarkably similar in these two species. The tissues display the same characteristics regardless of the histological stains used. Ameloblasts systematically present a cellular morphology associated with the secretion activity (highly elongated cells with displacement of the nucleus to the apical part) at the time of synthesis of the enameloid matrix. This morphology is temporary, and their morphology in more mature tooth buds may correspond to a maturation stage of the tissue. The mesenchymal cells (which thus give rise to the odontoblasts) present highly developed cytoplasmic processes, and well-differentiated odontoblasts are still visible in teeth at their latest developmental stages, including those that are already functional, which indicates that dentine production continues after the tooth becomes functional. Overall, the histological stains used here indicate a very strong presence of collagens in the dentine, which are also present in more limited quantities in the enameloid matrix. These observations have also been confirmed by electron microscope observation. Although no difference has been shown here between the dental development processes of the two taxa studied (which possess different microstructures), ultrastructural observations available in the literature (Kerebel *et al.*, 1977; Prostak *et al.*, 1990, Sasagawa, 1989; 2002) do illustrate the fact that while the microcrystals of selachimorphs and rajiforms seem to be fairly closely associated with the collagen fibers of the matrix, this is not the case for Myliobatiformes, which corresponds perfectly with the observations that have been made of the dental microstructure of these groups. In any case, the collagens shown on the basis of these histological markers do not seem to be synthesized in the ameloblasts. The role of these collagens thus remains difficult to establish with precision on the basis of these data, although they do seem to engage in a temporary secretory activity given their morphology.

Other pathways remain to be explored. Studies in the process of publication, based on the examination of patterns of gene expression of several genes involved in the synthesis of the extracellular matrix, may provide some initial answers. On one hand, unlike other vertebrates possessing enameloid, the ameloblasts of Chondrichthyes do not seem to synthesize collagen, but rather other proteins playing a crucial role in the production of mineralized tissues. On the other hand, if the nucleation of apatite microcrystals is necessary in order for the enameloid to become functional, the matrix must also be broken down, which is usually enabled by enzymes such as collagenases or other metalloproteinases (Sire *et al.*, 2009). Examining the patterns of expression of such genes should thus help us to specify the contribution of ameloblasts to the maturation of the enameloid. Given the morphology of ameloblasts in the early stages of odontogenesis, it is not impossible that they might also be involved in the synthesis of the matrix before its maturation and mineralization.

1.5. Dentin

Dentin makes up the largest part of the tooth crown in vertebrates. Most studies concerning the dental histology of chondrichthyans therefore focused initially on this tissue, due to its greater ease of observation and large volume compared to those of other dental tissues (e.g. Orvig, 1967; Peyer, 1968). It is a tissue quite similar to bone, rich in collagen, which shows significant histological variation and can sometimes prove quite difficult to distinguish from bone. This variation has led to the description of a very large number of types of dentin, the validity of which is sometimes doubtful, and to invalid phylogenetic hypotheses (Radinsky, 1961). For example, it was on the basis of dentin that the cochliodonts (an extinct group of chondrichthyans related to chimaera possessing grinding tooth plates) were linked to rays, and some authors even went so far as to propose classifications entirely based on dentine histology (Glikman, 1964; Thomasset, 1930). More broadly, dentin has also been used in the attempt to decipher the relationships existing between groups extremely distant from one another, such as placoderms, acanthodians and even some groups of agnathans.

In neoselachians, we mainly encounter two major types of dentin, osteodentin and orthodentin, which are subdivided into different categories according to the morphological variations appropriate to each of them (Cappetta, 2012). Thus, we can categorize the teeth of many Chondrichthyes into two main types of histological organization (Figure 1.8): orthodont, in which the tooth retains a pulp cavity surrounded by orthodentin and a base composed of osteodentin, and osteodont, in which the pulp cavity is secondarily filled in by osteodentin.

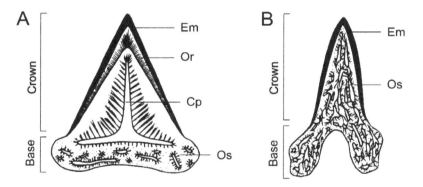

Figure 1.8. *Histological tooth types of neoselachians. A: orthodont (*Carcharhinus*). B: osteodont (*Lamna*). Em: enameloid; Cp: pulp cavity; Or: orthodentin; Os: osteodentin. Modified from Thomasset (1930)*

We refer interested readers to the highly detailed studies of Peyer, Ørvig and Poole for a complete review of all of the types of dentin described in the literature. In any case, though, as we have seen from the many questions remaining unanswered about the precise nature of enameloid, dentin is actually a much simpler tissue to understand in this context. It is recognized that dentin is a very consistent tissue in vertebrates. First seen in the dermal skeleton of certain ostracoderms (agnathan vertebrates), it is a collagen-rich tissue synthesized exclusively by odontoblasts, which has been demonstrated multiple times in various groups of vertebrates. While current data leave no doubt about the homology of dentin in vertebrates, this tissue does present a significant structural variation, the causes, as well as the functional implications, of which have yet to be clearly established. In any case, it seems clear today that it is of far less interest than enameloid in a phylogenetic and/or taxonomical context (Blazejowski, 2004).

1.6. Homology of enameloid tissues

The examination of questions pertaining to the developmental details of these tissues also question their homology with other mineralized tissues across vertebrates. Although we have seen that the homological argument concerning dentin is a robust one, this is not the case with enameloid. Even though the lack of homology with enamel seems well established, the relationships of the different kinds of enameloid with each other remain problematic. This type of tissue is seen particularly in many actinopterygians, where it is called acrodin (see Chapter 6 of this book), as well as in the larval stages of at least one species of amphibians, the Spanish ribbed newt (*Pleurodeles waltl*) (Assaraf-Weill *et al.*, 2014; Huysseune *et al.*, 2008; Kawasaki *et al.*, 2005; Kawasaki, 2009). From a morphological perspective, the microstructure of the enameloid of certain teleosteans also shows a striking resemblance to structures described in neoselachians, which has usually been interpreted as a convergence phenomenon (Reif, 1979; Shellis and Berkowitz, 1976). In fact, the two tissues show a number of differences in the compositions of their extracellular matrix, which is richer in collagen in actinopterygians than in Chondrichthyes, and in their patterns of mineralization (which begins at the enameloid/dentine junction in actinopterygians, while it is more diffuse in chondrichthyans) (Huysseune and Sire, 1998; Prostak and Skobe, 1986; Prostak *et al.*, 1993; Sasagawa, 1993; Shellis and Miles, 1974). The absence of enameloid in the teeth and scales in some of the earliest known Chondrichthyes, such as *Cladoselache*, *Holmesella*, *Xenancanthus* and *Adamantina*, as well as in some actinopterygians, including *Cheirolepis* and *Psarolepis* (although the latter may be a sarcopterygian), also points to a convergence of enameloid in Chondrichthyes and actinopterygians (Bendix-Almgreen, 1993, 1994; Dean, 1909; Orvig, 1966). On the contrary, acrodin is present in the teeth of *Andreolepis* (Janvier, 1978), one of the oldest known actinopterygians; however, this paleontological fact has been considered too anecdotal to constitute an argument in favor of the convergence of enameloid in these two groups (Gillis and Donoghue, 2007).

The potential presence of enameloid in some euchondrocephals has also been put forward as an argument in favor of rejecting the convergence hypothesis. The differences in the matrix and mineralization observed between chondrichthyans and actinopterygians would thus be the product of divergences within each of the two groups (Gillis and Donoghue, 2007). However, several problems remain; on one hand, the nature of the hypermineralized tissue present on the tooth surfaces of euchondrocephals is still subject to controversy. It could be pleromin, a tissue composed of whitlockite, a rarer form of calcium phosphate than hydroxyapatite (Ishiyama *et al.*, 1984, 1991), rather than true enameloid. However, it is difficult to rule one way or another in this question, considering the paucity of data currently available. That being said, some data generated by molecular biology have contributed a few new elements. The recent focus on genes in the family of SCCPs, some of which code for EMP, previously unknown in actinopterygians, in the genome of the spotted gar, provides support for the hypothesis that some of these genes may be involved in the formation of acrodin (Qu *et al.*, 2015). Finally, we would also point out that even though it is not part of the EMP, a gene called ODAM belonging to the SCCP family and coding for a protein involved in the maturation of hypermineralized tissues presents comparable patterns of expression in tetrapods and teleosteans (Kawasaki, 2009; 2013). These results have led to the hypothesis of a mineralization process shared within Osteichthyes and common to both enamel and acrodin. The probable absence of SCCP from the genome of chondrichthyans (Venkatesh *et al.*, 2014) theoretically excludes the sharing of such a process with osteichthyans, which means that the mineralization of the selachian enameloid also differs from that of the enameloid of actinopterygians from a molecular point of view. These data support the hypothesis that enameloid in the broader sense designates a group of tissues that are probably convergent in Chondrichthyes and Osteichthyes. However, it will be necessary to examine and compare the patterns of expression of several of the genes involved in the formation and mineralization of enameloid in multiple osteichthyans and chondrichthyans. Obtaining such data could also enable a better understanding of the evolutionary relationships existing between these tissues.

Paleozoic Sharks

2.1. Do all Paleozoic shark teeth possess enameloid?

The Paleozoic Era begins with the Cambrian Period (541 million years ago) and ends with the Permian Period (252 million years ago). The oldest fossil teeth that can be attributed with certainty to Chondrichthyes were discovered in the Lower Devonian of Spain and have been dated to approximately 418 million years ago (Botella *et al.*, 2009). They have been attributed to a species called *Leonodus carlsi*. The teeth have a bicuspid morphology, i.e. they possess two cusps of the same size, one at each end of the crown. Such morphology has no real equivalent in modern sharks (selachimorphs), in which the main cuspid is centrally positioned on the crown. Paleozoic forms are in reality phylogenetically distant from selachimorphs, and most of them belong to a set of groups called stem-chondrichthyans, which are often highly different from one another and appeared even before the separation of the lineages that led to the modern forms, the neoselachians (rays and selachimorphs) on the one hand, and the chimaera on the other hand (Ginter *et al.*, 2010). With these stem-chondrichthyans being very ancient and morphologically different from their modern relatives, it is difficult to decipher their phylogenetic relationships, all the more since they are mostly known from isolated teeth only. Moreover, the study of these teeth is all the more difficult since their morphology and microstructure have been altered during their long time buried in the Earth.

Despite its great age, researchers have managed to examine the microstructure of the enameloid on the teeth of *Leonodus carlsi* (Botella *et al.*, 2009). At the junction between the enameloid and the dentin, numerous dental tubules can be seen which penetrate the lower part of the enameloid. This relatively poorly defined boundary between enameloid and dentin is very different from what is seen at the dentin–enamel junction in tetrapod teeth. In the latter, odontoblasts do not participate in the making of enamel, which is carried out solely by ameloblasts (Kawasaki, 2009). Dentin and enamel are generated by two different types of cells, and there is no interpenetration between the two tissues. Conversely, it is possible that odontoblasts participate in association with ameloblasts in the making of enameloid, whether in *Leonodus* or in other fish possessing this type of hypermineralized tissue (see Chapter 1). This lack of segregation between cellular domains might also explain the relative intertwining of dentin and enameloid.

The enameloid microstructure of *Leonodus* has been shown to be of SCE type; it is composed of microcrystals that do not show any particular organization. Given the age of *Leonodus* teeth, it is tempting to think that an SCE represents the ancestral microstructure of the enameloid of selachian teeth, from which all other types of enameloid have developed (see following chapters).

Surprisingly, the teeth of other Paleozoic selachians, also considered very primitive and sharing the same bicuspid morphology as *Leonodus*, seem to lack enameloid. These include the teeth of *Aztecodus* and *Antarctilamna*, both coming from the top of the Middle Devonian (383 million years ago) of the Antarctic (Hampe and Long, 1999). However, another chondrichthyan called *Portalodus*, coming from the same geological layers and sharing similar dental morphology, shows SCE-type enameloid in which the microcrystals tend to amalgamate into bundles that lie perpendicular to the surface (Hampe and Long, 1999). Thus, it appears that the dental microstructures of these earliest Paleozoic selachians were remarkably diverse.

The teeth of *Mcmurdodus whitei* are, from a stratigraphic point of view, midway between those of *Leonodus* and those of *Aztecodus, Antarctilamna* and *Portalodus*. They come from the Early Devonian (393 million years ago) of the state of Queensland in Australia. These teeth are interesting on more than one front, being morphologically very different from the bicuspid ones of *Leonodus* and of the other Australian taxa mentioned in the last paragraph. In many regards, they are more similar to the dental morphology of modern sharks, such as the Hexanchiformes (a modern group that includes the cow sharks) and the Echinorhiniformes (bramble sharks). *Mcmurdodus* is therefore sometimes considered to be the oldest representative of the Selachimorph clade, which includes all the modern sharks but excludes skates and rays. Yet, selachimorphs are characterized by a highly complex enameloid microstructure (see Chapter 5). What is the case with *Mcmurdodus?* Though the presence of enameloid on its teeth has been demonstrated, their state of preservation does not allow a detailed examination of its structure. The characteristic complexity of selachimorphs, however, does not seem to be present (Burrow *et al.*, 2008).

The teeth of *Cooleyella* are more recent, dating from the Early Carboniferous Period (345 million years ago) to the Middle Permian Period (273 million years ago) and, like those of *Mcmurdodus*; they are sometimes considered as belonging to very primitive selachimorphs. However, preliminary studies have suggested that they did not possess enameloid (Duffin and Ward, 1983).

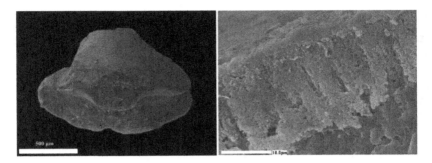

Figure 2.1. Cooleyella fordi *tooth from the Lower Carboniferous (340 mya) of Derbyshire, England. The tooth has been treated with HCl (diluted to 10%) for 5 seconds. To the left is an apical view of the tooth showing its broken tip, which enables us to see the enameloid in a cross-section. To the right is a detail of this cross-section showing SCE organized in poorly defined bundles oriented perpendicular to the surface of the tooth*

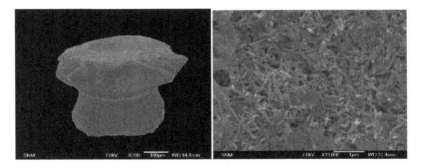

Figure 2.2. Ginteria fungiforma *tooth of the Lower Carboniferous of Russia (335 mya). The tooth has been treated with HCl (diluted to 3.7%) for 5 seconds. To the left is a detail of the enameloid surface showing SCE composed of randomly oriented microcrystals*

One of the authors of this book (GC), with the assistance of his Russian colleague Alexander Ivanov of the University of St. Petersburg, has been able to re-examine the microstructure of *Cooleyella* teeth as well as those of its close relative *Ginteria* from the Early Carboniferous Period. In both the cases, a layer of SCE-type enameloid was observed. In *Cooleyella*, the thickness of this tissue varies between

10 and 20 µm and its microcrystals show a clear tendency to be grouped into bundles perpendicular to the surface (Figure 2.1), a structure reminiscent of the teeth of *Portalodus*. However, the microcrystals do not show any preferred orientation within the bundles, whereas the detail of the microstructure of the bundles in *Portalodus* teeth has not yet been examined in detail. Conversely, no bundles have yet been observed in *Ginteria*, since only the surface of the tissue has been studied to date (Figure 2.2).

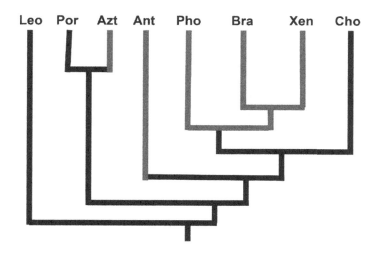

Figure 2.3. *Phylogenetic tree illustrating the phylogenetic relationships between basal selachians, in which the tooth enameloid microstructure has been successfully studied. Leo: Leonodus, Por: Portalodus, Azt: Aztecodus, Ant: Antarctilamna, Pho: Phoebodonts, Bra: Bransonelliformes, Xen: Xenacanths and Cho: rest of the Chondrichthyes. Red branches indicate the lines in which the teeth do not seem to possess enameloid. This tree is based on the phylogenetic hypotheses of Ginter, Hampe, and Duffin (2010). For a color version of this figure, see www.iste.co.uk/cuny/selachians.zip*

Another group of Paleozoic selachians is considered to have possessed teeth lacking enameloid. This is the clade that includes phoebodonts, bransonelliforms and xenacanths. These groups form a relatively diversified clade that includes several dozen species of which the earliest representatives, the phoebodonts, first appeared in the Middle Devonian Period (385 million years ago), and the last representatives, the xenacanths, lasted until the Late Triassic Period (210 million years ago). Phylogenetically speaking, these selachians are more primitive than *Mcmurdodus, Cooleyella* or *Ginteria*, but more derived than *Leonodus, Antarctilamna, Aztecodus* and *Portalodus* (Figure 2.3), with which they share a relatively similar dental morphology, which is considered primitive for Chondrichthyes (Ginter *et al.*, 2010).

Within this clade, the teeth of xenacanths have been the subject of the most histological studies, and the presence of enameloid on them is difficult to demonstrate. The outermost layer of the tooth crown has been interpreted, rather, as corresponding to a specific type of dentin called pallial dentin, or vitrodentin, though these terminologies are problematic, as noted in the previous chapter. Fluorescent optical microscopy studies suggest that this material is less dense than the underlying dentin, a characteristic incompatible with hypermineralized tissue such as enameloid (Hampe, 1991). Additional studies using electron microscopy have enabled the detail of this layer to be seen (Figure 2.4); its microstructure is fairly globular in appearance, which may be due to an artifact of preparation and/or fossilization, and microcrystals cannot be discerned within it. In dentin or bone, the degree of mineralization, generally approximately 70%, is not high enough to allow the precipitation of microcrystals exceeding 30 nm in their largest dimension (Berkovitz and Shellis, 2017).

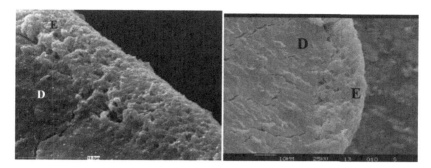

Figure 2.4. *Left: cross-section of enameloid at a break in an* Orthacanthus *sp. tooth from the Lower Permian of Buxières-les-Mines (Department of Allier, France). Right: break in a* Bransonella nebraskensis *tooth from the Lower Carboniferous of Derbyshire, England. D: dentin and E: enameloid-like tissue. Both teeth have been treated with HCl (diluted to 10%) for 5 seconds*

We have observed the same globular microstructure in a bransonelliform (Figure 2.4) as well as in a phoebodont (*Thrinacodus ferox* from the Lower Carboniferous of England). A similar "microstructure" has also been described in some chimaeras (Ishiyama *et al.*, 1991). The absence of enameloid in another phoebodont (*Phoebodus*) has been pointed out by Oliver Hampe, who has worked extensively on the dental microstructure in xenacanths, as well as by an Australian team (Hampe and Long, 1999; Turner and Burrow, 2011).

The tissue, which we have observed in *Orthacanthus, Bransonella* and *Thrinacodus,* with its globular structure and lack of clearly individualized microcrystals, therefore does not seem to fulfill the definition of an enameloid. However, it appears distinctly separate from the underlying dentin, and currently its nature remains a mystery. Is it a specialized type of dentin, or an enameloid that has been demineralized or altered by the fossilization process? Do ameloblasts participate in the making of this tissue, or is it solely the result of odontoblast activity? For now, we do not have the answers to these questions, but this relatively little-mineralized external covering of the tooth appears to be a characteristic (synapomorphy) of the clade encompassing phoebodonts, bransonelliforms and xenacanths.

Ultimately, the initial stages of enameloid evolution in chondrichthyans remain difficult to interpret. The oldest chondrichthyan teeth known to date possess SCE-type enameloid, suggesting that this was the ancestral condition for these animals. However, the detail of its microstructure remains difficult to specify. Rough bundles oriented perpendicularly to the tooth surface, as in *Portalodus* and *Cooleyella*, are sometimes seen, but most of the time it is difficult to obtain high enough resolution to make out such details. The distribution of these bundles is therefore not precisely understood. However, as we will see in the rest of this book, the presence of bundles is a characteristic found very often in tooth enameloids. Is it a part of their ancestral configuration, or did it appear at a somewhat later point in the evolution of the tissue?

In addition to this difficulty in understanding the primitive microstructure of enameloid, the propensity of this tissue to disappear should also be noted. This just raises the question of whether or not the oldest teeth in the fossil record represent the ancestral condition. Enameloid seems to be lacking in forms almost as ancient as *Leonodus*, like *Aztecodus* and *Antarctilamna*. If we take into account the small number of teeth of which the microstructure has been studied, how can we be sure that the presence of enameloid in the oldest teeth does not simply represent a sampling bias, and that teeth of the same age lacking enameloid are still not hidden somewhere in the fossil record? On the contrary, it is highly probable that the teeth resulted from the exaptation of dermal denticles into the jaws (the so-called "outside-in" theory; see Blais *et al.*, 2011; Smith and Hall, 1990; Witten *et al.*, 2014). Yet, all the dermal denticles studied to date, though relatively few in number, possess SCE-type enameloid, at least in their outer parts (Figure 2.5). This type of enameloid is also present in the dermal skeletons of ostracoderms, a group of Paleozoic vertebrates that lack jaws and are thus more basal than Chondrichthyes. It would therefore be logical to find this SCE in all of the earliest teeth.

In addition, the extreme age of these teeth and their small sizes are the limitations that make the characterization of their enameloid technically difficult. Thus far, only the teeth of phoebodonts, bransonelliforms and xenacanths have yielded enough information to make credible the hypothesis which proposes that the outer layer of their dentition does not fulfill the definition of an enameloid. However, we are not in a position where we can demonstrate that ameloblasts did not participate in its formation. In the cases of *Aztecodus* and *Antarctilamna*, the data obtained simply does not allow us to demonstrate the presence of such tissue, which may merely be indicative of our technical inability to observe it.

From a purely adaptative point of view, it is difficult to imagine why the enameloid of functional teeth, once in place, would disappear, and not once, but many times. This would lead to weakened mechanical resistance of the teeth. Perhaps a taphonomic explanation should be sought? Is it possible that this characteristic microstructure was erased by a secondary recrystallization of the teeth during their long stay in the sediments? Though this theory may be applicable to very ancient teeth such as those of *Aztecodus* and *Antarctilamna*, it seems unlikely for the teeth of more recent xenacanths. Enameloid can easily be seen on the teeth of similar age (see next section).

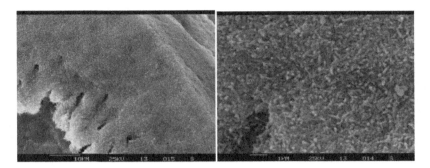

Figure 2.5. *Dermal denticle of a hybodont shark from the uppermost Triassic (208 mya) of the southwest of England, treated with HCl (diluted to 10%) for 5 seconds. Left: break in the enameloid showing compact SCE penetrated at the base by dentin tubules. Right: detail of SCE showing the absence of preferred microcrystal orientation*

In conclusion, the small number of teeth studied to date, separated by significant time intervals of 10 million years or more, and the sometimes insufficient resolution of current observation techniques, greatly limits our ability to understand the first stages of enameloid evolution in chondrichthyans. Though the theory, that the very first teeth possessed SCE-type enameloid, remains the most probable, the detail of its ancestral microstructure remains difficult to discern. Moreover, the mechanisms of the possible disappearance of this tissue in some lineages are not understood either. Only additional studies on a larger number of Paleozoic teeth and the improvement of observation techniques will enable us to overcome this impasse.

2.2. Ctenacanth sharks

We cannot help but be struck by the diversity of the structures observed within the enameloid of the oldest chondrichthyans as described in the previous section. This diversification continued during the second half of the

Paleozoic, as demonstrated by the study of the ctenacanth sharks. They appeared during the Late Devonian (383 mya) and, until recently, it was thought that they had disappeared during the Middle Permian (265 million years ago). The morphology of their teeth is different from the mostly bicuspid morphology of the first chondrichthyans, and is closer to that of modern sharks with a main cusp located at the center of the crown and flanked on each side by one or more pairs of secondary cusplets (Ginter *et al.*, 2010).

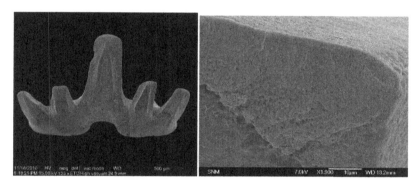

Figure 2.6. Glikmanius *cf.* myachkovensis *from the Middle Permian of Oman. Left: labial view of tooth. Right: cross-section of enameloid at the break point of a tooth treated with HCl (diluted to 3.7%) for 5 seconds, showing very compact SCE without bundles*

During her doctoral thesis, Martha Koot, under the supervision of one of this book's authors (GC), studied the microstructure of the enameloid in the teeth of a ctenacanth from the Middle Permian (266 mya) of Oman, belonging to the genus *Glikmanius*. The results of this study indicated the presence of compact and homogeneous SCE without bundles (Figure 2.6), reminiscent of what has been observed in *Leonodus* and dermal denticles. This microstructure has thus been interpreted as being an ancestral one, as might be expected for animals dating from the Paleozoic Era (Koot *et al.*, 2013).

At the same time, and without having been updated on each other's respective projects, another author of this book (GG) made a no less remarkable discovery: tiny ctenacanth teeth in the Lower Cretaceous (120 mya) of the south of France. This discovery extended their temporal distribution by 145 million years, an idea that was met with some skepticism. In order to eliminate the possibility that these teeth might belong to some atypical selachimorph, Guillaume and his colleagues proceeded with a study of their microstructure (Guinot *et al.*, 2013). As noted in Chapter 1, selachimorphs are characterized by a complex and easily recognizable microstructure. To their great surprise, they observed an enameloid composed of bundles lying parallel both to one another and to the surface of the tooth. Such a structure was heretofore known to exist only in selachimorphs.

These bundles are very different from the ones observed in *Portalodus* and *Cooleyella*; this is partly due to their orientation not being perpendicular to the surface as in these two genera, but above all it is due to their structure. The bundles seen in *Cooleyella* actually represent only separate bundles of microcrystals, but inside each bundle, the microcrystals do not really show a preferred orientation. Conversely, in the bundles of the Cretaceous ctenacanth teeth observed by Guillaume and his colleagues, the microcrystals are all oriented parallel to the axis of the bundles, thus forming a much more compact structure.

In order to be absolutely sure, Guillaume and his team decided to study the microstructure of a morphologically undisputable ctenacanth tooth: *Neosaivodus flagstaffensis*, described the previous year by John-Paul Hodnett and his team and dating from the Middle Permian of Arizona. They discovered a microstructure that was similar in every point to that of their Cretaceous specimens, a result contradictory to the one obtained by Gilles and Martha Koot!

How can this contradiction be explained? Ctenacanths form a group in which the phylogenetic relationships are difficult to decipher from the scattered teeth that have been found. Yet, it would seem that *Glikmanius* and *Neosaivodus* in fact belong to two different families (Hodnett *et al.*, 2012), families that could be characterized by, among other things, the microstructure of their enameloid. Heslerodids, to which *Glikmanius* belongs, would have retained a primitive microstructure, whereas ctenacanthids, which include *Neosaivodus*, would have developed a more complex structure. For now, this remains a theory that can only be confirmed by studying a larger number of ctenacanth teeth; however, it illustrates the potential of enameloid microstructure as a tool enabling us to better understand the evolution of chondrichthyans. We will further illustrate this potential in the following chapters.

The work of Guillaume and his team has thus made it possible to demonstrate that a group of sharks survived, well hidden in the depths of the ocean, between 265 and 120 million years ago. The teeth studied also have the particular feature of having been discovered in sediments indicating a deep water marine environment. The history of the ctenacanths is therefore reminiscent of that of the coelacanths, which were believed to have vanished after the extinction event at the end of the Cretaceous Period, until living specimens were found in 1938 in the depths of the Indian Ocean. Thus, we have an illustration of the role of the deepest parts of the oceans as "refuge areas". They are indeed characterized by a highly stable environment, far from the climatic disturbances affecting surface waters, which may lead to the extinction of very specialized lineages.

As we will see in Chapter 5, the tooth enameloid of ctenacanths, despite the presence in some genera of highly structured bundles, does not have the complexity that we see today in modern selachimorphs. Still, during the course of the Paleozoic Era, we see a significant complexification of this tissue, and recent discoveries have given us cause to believe that we are still far from having taken the full measure of the evolutionary potential of this tissue.

3

Hybodont Sharks

Hybodont sharks constitute the sister group of the neoselachians (Maisey *et al.*, 2004), which include all living elasmobranchs and encompass two major clades: batomorphs (skates and rays) and selachimorphs (modern sharks) (Douady *et al.*, 2003). The overall morphology of hybodont sharks is quite similar to that of the latter, but is distinguished by the absence of calcified vertebral centra, the possession of dorsal spines with a concave posterior wall and the presence of cephalic hook-shaped spines in males, which probably helped to grip the female during mating (Cuny and Bénéteau, 2013). The first hybodont fossils date back to the Late Devonian and are represented only by isolated teeth attributed to the species *Roongodus phijani* in Iran and *"Lissodus" brousclaudiae* and *"Lissodus" tursuae* in Belgium (Hodnett *et al.*, 2012). The earliest more or less complete fossils date back to the Early Carboniferous Period, particularly *Onychoselache traquairi*, discovered in Scotland (Ginter *et al.*, 2010). The hybodonts became extinct at the end of the Maastrichtian Age, at the same time as the non-avian dinosaurs (Cappetta, 2012). Thus, in a sense, they represent an intermediary between the Paleozoic chondrichthyans, which we discussed in the last chapter, and modern elasmobranchs.

3.1. Enameloid microstructure in hybodont teeth

Somewhat paradoxically, the microstructure of hybodont tooth enameloid has not been extensively studied, as most of the research, as we will see, has focused on neoselachian enameloid, which is thought to be both more characteristic and more complex. Indeed, the interest in hybodont enameloid microstructure is mainly due to the status of hybodonts as the sister group of neoselachians; in fact, it is often difficult to distinguish between the earliest representatives of the two sister lineages just after their divergence, when, generally speaking, the taxa have not yet had the time to develop clear synapomorphies (derived characteristics) and share many more ancestral characteristics (symplesiomorphies) than derived ones. This is a well-known problem in cladistics, which often leads to the establishment of paraphyletic stem groups. The problem is further exacerbated when, as in the case of chondrichthyans, most of the fossils are highly fragmentary and represent only a small part of the animal, in this case mostly teeth. For example, the teeth of some primitive selachimorphs, such as synechodontiforms, are morphologically very similar to those of hybodonts (Figure 3.1). Therefore, the study of enameloid microstructure has rapidly emerged as a promising method of distinguishing between the teeth belonging to these two groups.

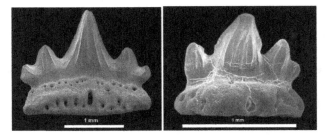

Figure 3.1. *Comparison of a tooth of* Parvodus celsucuspus, *a Lower Cretaceous hybodont from Charente, France (left), and a tooth of* Rhomphaiodon minor, *a primitive selachimorph from the Upper Triassic of England (right). Note the similarities in the overall morphology, particularly at the crown. Both teeth are shown in lingual view*

In his work in the early 1970s, Wolf-Ernst Reif discovered very different microstructures within these two clades (Preuschoft *et al.*, 1974; Reif, 1973). Neoselachians are characterized by a complex three-layered enameloid (we will return to this point in detail in the following chapters), whereas hybodonts possess a much simpler enameloid formed of a single homogeneous layer composed of randomly oriented microcrystals (the SCE we saw in *Leonodus* in the last chapter). With hybodonts representing the sister group of neoselachians, this structure was considered to be ancestral, and most research was focused on studying the derived condition of neoselachians. Consequently, most hybodont teeth were studied only via the surface acid etching method, with the general idea being that in an event of doubt about whether an isolated tooth belonged to this (or that) clade, an etching would be sufficient to show (or not) the complex structure characteristic of neoselachians. In fact, this method was aimed only at quickly identifying possible primitive neoselachian teeth, rather than at studying in detail the evolution of enameloid in hybodont teeth.

There is a classic example that shows the difficulty of distinguishing between these two types of teeth. In 1836, the Swiss paleontologist Louis Agassiz established the genus *Hybodus* on the basis of isolated dorsal spines and teeth discovered in the Upper Triassic System (Rhaetian Stage) in England (Agassiz, 1836). The species *H. minor* is one of the three species assigned to this genus by Louis Agassiz, a genus representing the archetypical hybodont shark. However, it was only in 2005 that a microstructural analysis of *Hybodus minor* enameloid revealed that this genus was not a hybodont, but a primitive neoselachian. As a result, they were assigned to the genus *Rhomphaiodon* (Figure 3.1) (Cuny and Risnes, 2005). The correction of the phylogenetic position of these teeth at the ordinal level, then, took 169 years – effectively illustrating the difficulties that may be encountered by paleontologists in studying the evolution of fossil sharks. It was finally realized that the teeth and dorsal spines (the latter of which belong to a hybodont), attributed to the same species by Louis Agassiz because they were found together at the same sites, in fact belonged to two completely different sharks. This tells us that enameloid microstructure is a tool enabling us to better understand not only the phylogenetic relationships of isolated teeth but also the composition of fossil vertebrate assemblages composed solely of isolated parts, as is the case in the European Late Triassic. Indeed, it is amusing to think that if Agassiz had based the genus *Hybodus* on the species *H. minor* rather than the species *H. reticulatus* (the former was named on page 183 of volume 3 of his monumental work *Recherches sur les poissons fossiles* (Research on Fossil Fish), while the latter was named on page 180), the whole group of hybodonts as currently defined would have been based on an animal that was not one of them!

However, recent studies have shown that the microstructure of the enameloid in hybodont teeth, although relatively homogeneous, is more complex than previously thought (Enault *et al.*, 2015). Two units can often be seen in it, although they are not always clearly separated. The outer unit is generally dense and homogeneous and is composed of randomly oriented microcrystals, a characteristic that is particularly visible when the tissue surface is observed under acid etching. Thus, it belongs to the category of SCE (single-crystallite enameloids) in the terminology we introduced in 2015. Indeed, this outer unit corresponds on all points to the idea put forward by Wolf-Ernst Reif concerning the whole of the tissue in hybodonts. On the contrary, the internal unit, when present, appears more complex, because the microcrystals making up the tissue show a clear tendency to be grouped in bundles oriented perpendicular to the surface of the enameloid. The presence of bundles places this unit within the BCE (bundled crystallite enameloids), and more precisely the RBE (radial bundled enameloids), as they lie perpendicular to the tooth surface (Figure 3.2). It is important to note, however, that the bundles in hybodonts are much more disorganized than those in modern sharks; in the latter, all of the microcrystals are oriented parallel to one another and to the axis of the bundles, which is not the case in hybodonts, in which the microcrystals show significant variations in orientation. One of the consequences of this freer arrangement, combined with the fact that the bundles are oriented perpendicular to the surface, is that the latter are virtually undetectable in studies of the tooth surface. To view them, a cross-section of the tooth being studied must be made. This partly explains why the presence of bundles within hybodont enameloid was only recently shown. It is noteworthy that hybodont RBE gives the impression that the main characteristic of their enameloid is its overall perpendicular orientation to the surface, rather than parallel orientation, as seen in certain ctenacanths in the previous chapter, and which is also found in neoselachians.

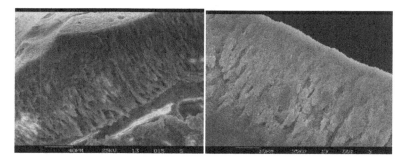

Figure 3.2. *Left: cross-section of the enameloid of an Acrodus* spitzbergensis *tooth from the Middle Triassic of Nevada, treated for 35 s with 10% dilute HCl. Right: cross-section of the enameloid of a* Lissodus minimus *tooth from the Upper Triassic of England, treated for 30 s with 10% dilute HCl. In both cases, the SCE is thinner than the BCE, the bundles of which are more highly individualized in* A. spitzbergensis *than in* L. minimus

A similar arrangement can also be seen in the enameloid of the large buttonlike dermal denticles found in hybodonts (Figure 3.3), which suggests that it is a plesiomorphic condition (Enault *et al.*, 2015). The dermal denticles covering the bodies of chondrichthyans have an overall structure similar on all points to that of the teeth with which they probably share a common evolutionary origin (Blais *et al.*, 2011). Unfortunately, the small size of these dermal denticles – less than half a millimeter – makes studying the microstructure of their enameloid technically complicated or even impossible. The "large" "buttonlike" dermal denticles of hybodonts represent one of the rare exceptions allowing this type of study.

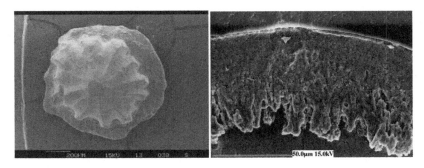

Figure 3.3. *Left: apical view of a "button"-shaped dermal denticle of a hybodont from the Upper Triassic of eastern France. Right: cross-section of the enameloid of a similar dermal denticle from the Upper Triassic of England, treated for 5 s with 10% dilute HCl*

However, some authors have recently pointed out that hybodont RBE may in fact correspond to an artifact due to the penetration of dentine tubules into the inner part of the enameloid, which would have "forced" the enameloid into an arrangement of radial bundles (Manzanares *et al.*, 2016); this is an idea that must be explored and tested further.

3.2. Development of specialized tooth structures and associated changes in the enameloid microstructure

Hybodonts possessing small and little-specialized teeth, with low crowns and little-developed cusps, such as *Lissodus*, for example, show a microstructure that corresponds to the one described above, with a more or less well-developed SCE and RBE (Figure 3.2) (Enault *et al.*, 2015; however, see Manzanares *et al.*, 2016, for a different interpretation). However, this general organization may have changed significantly with the appearance of more specialized tooth structures.

Starting at the end of the Jurassic, several genera (*Egertonodus, Hybodus, Meristonoides, Planohybodus, etc.*) developed tearing-type dentitions characterized by a crown with a high central cusp flanked by smaller lateral cusplets. The enameloid in such teeth shows a clear tendency toward homogenization. Its fluorapatite microcrystals show fewer variations in orientation, with a layout that is mainly perpendicular to the tooth surface, and the bundles are less distinctly separate from one another. Thus, the enameloid appears more compact and there is less differentiation between SCE and RBE. One probable explanation for this microstructural change lies in the fact that high crowns are subject to greater tensile stresses than low crowns. The compaction of the tissue would enable better resistance to this type of constraint. However, unlike what is seen in modern sharks, no specific structure is present to strengthen the cutting edges or ornamentation ridges of the teeth. All that can be seen at this level is a simple thickening of the tissue.

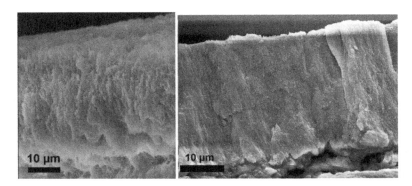

Figure 3.4. *Left: cross-section of the enameloid of a* Thaiodus ruchae *tooth from the Lower Cretaceous of Thailand, treated for 95 s with 10% dilute HCl. Right: cross-section of the enameloid of a* Priohybodus arambourgi *tooth from the Lower Cretaceous of Tunisia, treated for 5 s with 10% dilute HCl*

We find the same type of modification in the rare hybodonts to have developed a cutting dentition with serrated teeth (*Pororhiza, Mukdahanodus, Thaiodus*) (Duffin and Cuny, 2008). However, the crown of these teeth, unlike what is seen in the great white shark, for example, remains very low, which seems to imply that the resistance of hybodont enameloid to tensile stresses is relatively weak overall. There is a single exception, *Priohybodus,* in which we see high triangular teeth with serrated edges similar to what can be found in some modern sharks (Duffin, 2001). However, the homogenization and compaction of its enameloid are quite unlike what has been observed in every other hybodont. The microcrystals are all aligned perpendicular to the surface and the bundles, and thus the RBE have completely vanished (Figure 3.4). The appearance of high cutting-type teeth seems, therefore, to correlate with a drastic re-organization of the microcrystals forming the enameloid, which certainly explains the rarity of this type of teeth in hybodonts. This theory is corroborated by the fact that *Carcharopsis,* one of the rare Paleozoic selachians to have cutting teeth similar to those of *Priohybodus,* shows very similar compaction, although to a lesser extent, in its tooth enameloid (Duffin and Cuny, 2008). More generally, the fact that the number of hybodonts possessing cutting teeth is much lower than what we see today in modern sharks is undoubtedly related, at least in part, to the original microstructure of their enameloid.

However, the distribution of this type of teeth in hybodonts has yet to be explained. We do not find them before the Late Jurassic, even though the group first appears in the Devonian, and they only appear in animals which spent most, if not all, of their lives in freshwater lakes and rivers. One of our colleagues, Lars Stemmerik of the Natural History Museum of Denmark, suggested once that this might correspond to the occupation of a particular ecological niche: that of scavengers feeding on the carcasses of dinosaurs drowned when river levels rose. Why not? The development of this type of dentition would have greatly facilitated the carving up of this type of food. However, it goes without saying that the diet of these sharks was not limited to a single type of prey!

Conversely, numerous hybodonts, including *Asteracanthus* and *Heteroptychodus,* developed grinding-type dentition characterized by large, flattened teeth (Cappetta, 2012). These teeth contain SCE and BCE, but with significant changes in the BCE (Enault *et al.*, 2015). The boundary between the dentin and the lower edge of the enameloid is generally less clear than that in other forms. In particular, we can see dentine tubules penetrating deep into the enameloid (Figure 3.5). Between these tubules, the bundles constituting the inner unit lose their orientation largely perpendicular to the surface and show a tendency to become entangled (Figure 3.4). Thus, RBE is replaced by TBE. It should be noted that the appearance of this TBE seems to contradict the idea that these bundles are only artifacts due to the presence of dentin tubules, which are automatically oriented perpendicular to the surface. It is also noteworthy that the bundles tend to be more organized in large teeth than in small ones, suggesting that size has an impact on the apparent complexity of the microstructure. The thickness of the SCE varies from one genus to another, but does not show any significant changes.

Unlike teeth with high crowns, large, flat grinding teeth are not subjected to strong tensile stresses, but rather to compressive stresses. The entanglement of the bundles in the TBE and the presence of dentine tubules may in this case represent structures meant to dissipate these types of stresses, although this hypothesis would need to be tested more formally. Much remains to be done in the study of the mechanical characteristics of the different types of enameloid encountered in "fish" in general. It is important to note, however, that a similar microstructure is observed in the only modern selachimorph to have developed large grinding teeth, *Heterodontus* (see Chapter 5). This suggests a convergence phenomenon in the evolution of the tissue due to specific mechanical characteristics imposed by the durophagous diet of these animals.

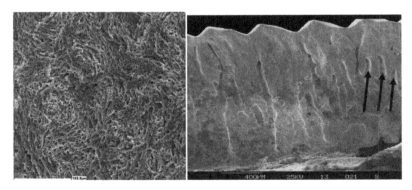

Figure 3.5. *Left: enameloid surface of an* Asteracanthus ornatissimus *tooth from the Upper Jurassic of France, treated for 35 s with 10% dilute HCl, showing organization in entangled bundles (TBE). Right: cross-section of a* Heteroptychodus steinmanni *tooth from the Lower Cretaceous of Thailand, treated for 95 s with 10% dilute HCl. The black arrows indicate dentin tubules penetrating the enameloid*

3.3. The mysterious *Ptychodus*

Ptychodus is in many ways one of the most mysterious sharks of the Cretaceous Period. It appeared at the end of the Early Cretaceous 110 million years ago and disappeared before the end of the Cretaceous, in the Campanian, approximately 70 million years ago. During the intervening period, we found them in abundance in every ocean on the planet and more than 20 species have been described to date (Cappetta, 2012), all of which correspond to a large animal between 2 and 10 m long, possessing an array of approximately 600 large grinding teeth (Shimada *et al.*, 2010). The size of the animal and the shape of its teeth suggest that it specialized in hunting ammonites and inoceramid bivalves, which were common bivalves during the Cretaceous Period. This prey could reach 2 m in diameter for the largest ammonites (Lewy, 2002) and 1.80 m in length for the inoceramid bivalves (Henriksen, 2008), explaining the gigantic size of *Ptychodus*.

Like most cartilaginous fish, *Ptychodus* is known mainly from its isolated teeth; however, these teeth show a highly unusual morphology, so much so that it has been a difficult matter to classify this shark. When he created the genus in 1839, Louis Agassiz considered it a hybodont, but its teeth possessed roots firmly attached to the crown, a neoselachian characteristic absent in hybodonts. Agassiz's interpretation was called into question quickly and *Ptychodus* began to be considered as a ray (Woodward, 1889). However, a fossilized jaw fragment indicates that these animals had a very elongated jaw, an unlikely possibility for a ray. The hyostylic articulation of their jaw, which is connected to the skull solely by means of hyomandibular cartilages, makes indeed the appearance of this morphology mechanically difficult.

In 1916, Mario Canavari's description of a calcified vertebral centrum associated with *Ptychodus* teeth in the Upper Cretaceous of Italy brought more credence to the idea that it was indeed a neoselachian (Canavari, 1916). Other similar discoveries were subsequently made in England, America and even the Antarctic (Stewart, 1980; Welton and Zinmeister, 1980). However, it should be noted that, in all these cases, these were simply fossils found together, rather than fossils found in connection. Therefore, we cannot dismiss the possibility that these associations were made of remains of different sharks, which is all the more plausible because *Ptychodus* teeth are very common in the fossil record. Neither can we dismiss the idea that calcified vertebral centra appeared secondarily via convergence in large hybodont sharks, in order to optimize their locomotion, partly because of the weight of their tooth set.

Ptychodus appears, then, to be an opportunity to test the effectiveness of the enameloid microstructure at resolving the phylogenetic relationships of sharks known only from teeth with unusual morphology. Of the teeth we have been able to study, its enameloid looks extremely similar to that of *Asteracanthus,* with an SCE, a TBE and dentine tubules penetrating the TBE (Figure 3.6). The TBE bundles are composed of microcrystals showing large variations in orientation. However, the teeth we studied showed a very flat morphology. A team led by Professor Brian L. Hoffman studied teeth with higher crowns and showed enameloid with an SCE, a PBE with a very pronounced radial component and a TBE (Hoffman *et al.*, 2016). Such a structure strongly suggests that *Ptychodus* is in fact a neoselachian. The simplest structure observed in the flattest teeth of other species would thus be only a secondary modification, having to do with a more extreme adaptation to a grinding function, convergent with the one observed in the hybodont *Asteracanthus* and the modern neoselachian *Heterodontus*. This is a theory that will need to be explored further in the years to come.

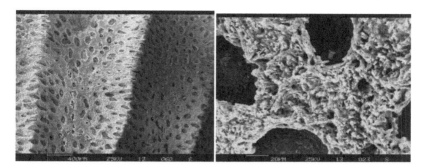

Figure 3.6. *Enameloid microstructure of* Ptychodus *sp. tooth with a highly flattened morphology from the Cenomanian (Middle Cretaceous) of Boulonnais (northern France). Left: surface of a tooth treated for 33 min with 10% dilute HCl showing numerous dentin tubules penetrating the lower part of the enameloid. Right: detail of the surface of the same tooth treated for 13 min with 10% dilute HCl, showing TBE between the dentine tubules*

3.4. Conclusion

The enameloid of the teeth and large dermal denticles of hybodonts is composed of an external SCE and an internal RBE. The best hypothesis is that this represents the plesiomorphic structure for this clade. However, this structure shows significant variations that often seem to correlate to the morphology and/or function of the teeth. The RBE has a tendency to resorb or vanish completely in forms possessing teeth with a high crown, whereas it turns into a TBE in hybodonts possessing large grinding teeth. In any case, the bundles remain relatively poorly defined and their microcrystals are never aligned along the axis of these bundles. The structure of the SCE is more constant, with the exception of the alignment of its microcrystals perpendicular to the surface as observed in *Priohybodus*.

Enameloid Microstructure in Rays

4.1. The phylogenetic position of skates and rays within the neoselachians

Skates and rays, also referred to as batomorphs, represent the most diverse group of modern chondrichthyans, with 633 known species (Last *et al.*, 2016). They show a wide range of morphological and ecological preferences and are found in almost all of the ecological niches occupied by modern sharks (Compagno, 1990; McEachran and Dunn, 1998). With the latter, they form the group of neoselachians (see Figure 1.2, Chapter 1); although the monophyly of modern sharks (selachimorphs) is well established today in terms of both molecular and morphological characteristics (Aschliman *et al.*, 2012; Cappetta, 2012; Delsate and Candoni, 2001; Douady *et al.*, 2003; Maisey *et al.*, 2004; Underwood, 2006), this has not always been the case. In fact, a very popular hypothesis in the late 1990s claimed that skates and rays were a highly derived group of sharks, related to forms such as the Squatiniformes (angel sharks) and the Pristiophoriformes (sawsharks) (de Carvalho, 1996; Shirai, 1996). These mostly benthic groups do share numerous morphological characteristics with skates and rays, but this hypothesis was definitively refuted by the first molecular phylogenies, the results of which confirmed the initial morphological analyses that skates and rays constitute the sister group of the selachimorphs (Douady *et al.*, 2003).

However, their evolutionary origins remain partially unclear. The first evidences of batomorphs in the fossil record date back to the Lower Jurassic, about 100 million years after the first-known neoselachians (Cappetta, 2012; Underwood, 2006). One of the authors of this book (GC) hypothesized that some neoselachians with uncertain affinities, such as *Pseudodalatias* (Triassic), *Vallisia* (Triassic), *Doratodus* (Triassic) or *Cooleyella* (Carboniferous), might represent the first-known batomorphs (Cuny *et al.*, 2009). In any case, it seems clear that the first batomorphs are indistinguishable from the selachimorphs on the basis of dental morphology alone. The characteristic dental morphology of rays does not seem to have been acquired before the Jurassic Period.

4.2. Historical context

We have seen in the previous chapters how the study of enameloid microstructure can, in some cases, help in differentiating neoselachian teeth from those of other chondrichthyans. Despite their extreme diversity, skates and rays have somehow fallen through the net, and their dental microstructure has remained virtually unknown until very recently (Enault *et al.*, 2015; Manzanares *et al.*, 2016). This apparent lack of interest in their dental histology can be largely explained by their long-unresolved phylogenetic relationships (interpretations of skates and rays as a group of highly derived sharks) and certainly by their small teeth as a general rule. The few studies on their dental histology, however, seemed to indicate the absence of the famous "triple-layered" enameloid, discussed in the first chapter of this book, and the possession of only a single-crystallite enameloid (SCE) (Reif, 1977; Thies, 1982). Yet this feature does not seem to have aroused any particular interest, and was interpreted in the end as an adaptation to a mostly durophagous diet (hard food sources composed of crustaceans and shelled mollusks) (Preschoft *et al.*, 1974, Botella *et al.*, 2009). Reconstructing the diet of a hypothetical batomorph as durophagous appears indeed to be the most parsimonious, but such an affirmation remains overly simplistic and fails to take into account the diverse range of dietary strategies known to exist among living skates and rays (Dean *et al.*, 2007).

However, renewed interest in this question, which was responsible in part for the collaboration among the three authors of this book, has enabled the demonstration of a significant organizational diversity within batomorph enameloid.

4.3. Batomorph enameloid: diversity and evolution

The results of the study of dental microstructures seen in skates and rays in recent years have revealed a much greater diversity than was suggested by the few descriptions available before this work (Enault *et al.*, 2015).

The oldest batomorph taxa whose enameloid microstructures have been examined here are *Toarcibatis elongata* and *Cristabatis crescentiformis*, two Archaeobatidae from the Early Jurassic (Figure 4.1), which already show a complex enameloid characterized by an apex composed of a SCE, that is, of non-bundled, largely homogeneous microcrystals. On the contrary, these microcrystals are grouped into short bundles that show little individualization and are parallel to the tooth surface on at least part of the crown, which is somewhat reminiscent of the organization seen in selachimorphs. Conversely, in areas farther from the tooth apex, these bundles show a greater tendency to be oriented perpendicular to the occlusal surface.

Belemnobatis sp., an early Cretaceous rhinobatid, possesses a microstructure similar to that of Archaeobatidae (Figure 4.2). However, contradictory results were obtained by one of us (GC) in the study of a different species of *Belemnobatis* (*B. aominensis*), with the latter being characterized by a single SCE (Cuny *et al.*, 2009). These results, based on two different approaches (section studies in *Belemnobatis* sp. vs. surface studies in *B. aominensis*), effectively illustrate the difficulties in observing poorly individualized bundles on surface treatments, but may also suggest the existence of a wide range of microstructures in genera with a lengthy stratigraphic distribution. Indeed, similar observations have been made in selachimorphs, with taxa such as *Synechodus* and *Paraorthacodus* (Guinot and Cappetta, 2011).

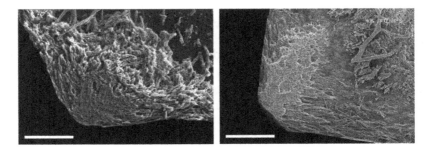

Figure 4.1. *Microstructures (cross-section) of* Toarcibatis elongata *enameloid (left) and* Cristabatis crescentiformis *(right). The scale bar is 20 µm. Both sections have been treated for 5 s with 10% dilute HCl*

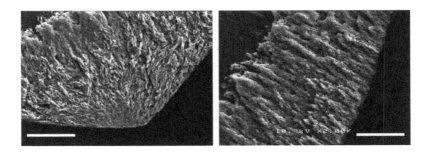

Figure 4.2. *Microstructure of* Belemnobatis *sp. in cross-section. Left: Detail of the transverse crest. The scale bar is 20 µm. Right: Detail of enameloid on the lingual face of the tooth. The scale bar is 15 µm. The sections have been treated for 5 s with 10% dilute HCl*

Much more complex microstructures are also known to exist in other taxa, particularly in the genera *Ptychotrygon* and *Parapalaeobates,* two fossil Rajiformes from the Cretaceous. These forms display a very thick enameloid with a complex, well-structured microstructure (Figure 4.3). In these taxa, the RBE is the most developed unit of their BCE. The SCE is also easily seen on the periphery of the enameloid layer. However, the possible presence of a PBE has not been clearly established, although bundles parallel to the occlusal surface can be seen, being completely enclosed within the RBE (Figure 4.3, white arrow).

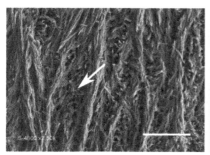 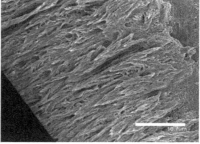

Figure 4.3. *Microstructure (cross-section) of enameloid from two Cretaceous Rajiformes. Left:* Parapalaeobates *cf.* atlanticus. *The white arrow indicates the presence of bundles that appear parallel to the occlusal surface of the tooth (because the tooth has been cross-sectioned, the bundles appear perpendicular to the surface of the photo), enclosed within the RBE. The scale bar is 12 µm. Right:* Ptychotrygon *sp. The scale bar is 17 µm. Both sections have been treated for 5 s with 10% dilute HCl*

In striking contrast to this complexity, other taxa, such as the genus *Hypsobatis*, also a Cretaceous rajiform, display only a SCE, completely lacking differentiation (Figure 4.4).

This lack of organization is somewhat reminiscent of the earliest descriptions of the dental histology of batomorphs (Preuschoft *et al.*, 1974; Thies, 1982). Similar morphologies are observed in myliobatiforms, illustrated here by the genus *Rhombodus*, a taxon of durophagous ray from the Early

Cretaceous (Figure 4.5). The microstructure of these forms seems systematically to show only a SCE. The case of the Myliobatiformes, a group of rays with highly specialized dietary habits, will be discussed in more detail in the next section, due to the particular interest it holds.

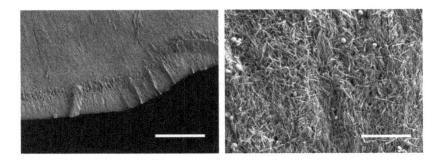

Figure 4.4. *Microstructure (section) of* Hypsobatis weileri *enameloid in cross-section. Left: Overall view of the enameloid layer. The scale bar is 120 μm. Right: Detail of SCE. The scale bar is 3 μm. The sections have been treated for 5 s with 10% dilute HCl*

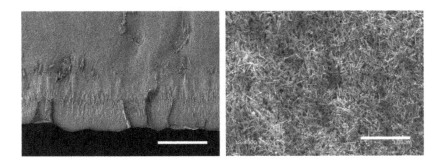

Figure 4.5. *Microstructure (section) of* Rhombodus binkhorsti *enameloid in cross-section. Left: Overall view of the enameloid layer. The scale bar is 200 μm. Right: Detail of SCE. The scale bar is 3 μm. Both sections have been treated for 5 s with 10% dilute HCl*

The largest sampling effort to date involves Rajiformes as well as Myliobatiformes, as these two orders show the greatest specific diversity among batomorphs. The situation within the Torpediniformes, which include electric rays, is less clear. To our knowledge, only one taxon has been studied to date: *Torpedo marmorata*, the marbled electric ray, a modern species living in the Atlantic and Mediterranean. The teeth of this taxon appear to show non-differentiated SCE enameloid; however, the observation of their dental histology is difficult due to the extremely small size of their teeth. In addition, the presence of organic material residue often makes it difficult to observe the crystalline microstructure of modern teeth (Enault, Adnet and Cappetta, 2013). In any case, data concerning a single taxon do not allow generalizations to be made for a whole group. The dental microstructure of electric rays thus remains largely unknown to date, as a more complete sampling, one that would notably include the most ancient forms known, such as *Eotorpedo* from the Early Palaeocene, remains to be conducted.

At any rate, it is clear that the earliest batomorphs identified as such already show a complex and highly differentiated microstructure, though with a less structured organization than what is known to exist in selachimorphs. This type of organization (BCE with well-developed RBE and/or TBE, but no PBE associated with an SCE), which is characteristic of many Rajiformes and somewhat reminiscent of the dental microstructure of some hybodonts, seems to constitute an ancestral arrangement in neoselachians. The existence of very simplified microstructures (made of SCE only) in some Rajiformes remains difficult to explain; however, they remind us that the boundaries of the group are still not well defined. In any case, this type of microstructure seems rarer. In general, regardless of the type of enameloid seen in the batomorphs examined, the junction between the dentin and the enameloid layer is often ill-defined and frequently shows dentinal tubules sunken into the enameloid layer. In conclusion, the Rajiformes, which grew especially diverse during the Mesozoic, present the most variable microstructure in terms of complexity and organization seen among the batomorphs. Conversely, the Cenozoic is marked by the rapid radiation of the Myliobatiformes, which will be the subject of the next section, and in which the enameloid microstructure is secondarily highly simplified.

4.4. Durophagy and other trophic specializations

As reiterated in the first part of this chapter, modern rays show a wide variety of feeding habits, and the same is certainly true for the fossil representatives of the group, judging by the wide morphological diversity of their dentition. However, the scarce early data available on the atypical dental microstructure of batomorphs were initially interpreted in an adaptive context (justified by a diet considered to be principally durophagous), which does not reflect the ecology of the group. Yet this interpretation of their dental microstructure – based, moreover, on an extremely small sampling – has remained firmly rooted in the literature (Cuny and Benton, 1999; Cuny and Risnes, 2005; Gillis and Donoghue, 2007; Underwood, 2006).

The best example of the diversity of diets existing in batomorphs is offered by the clade of Myliobatiformes, which notably include the manta ray (*Manta birostris*). This group is particularly interesting for the study of the impact of phylogeny and diet on dentition, as it brings together families that are closely related but present diametrically opposed trophic strategies (Figure 4.6), including durophagous (Myliobatidae, Rhinopteridae) and planktivorous (Mobulidae) forms (Adnet *et al.*, 2012). Thus, we may suppose that the evolution of their dental microstructure is liable to reflect both the mechanical stresses linked to their diet (or the absence of stresses, in the case of a planktivorous diet) and a phylogenetic signal related to their common evolutionary history.

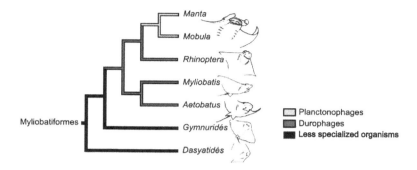

Figure 4.6. *Phylogeny and diet of the Myliobatiformes, from (Aschliman et al., 2012; Dean et al., 2007; Dunn et al., 2003). Figure modified from (Adnet et al., 2012). For a color version of this figure, see www.iste.co.uk/cuny/selachians.zip*

The existence of such marked specializations makes them an ideal group for evaluating the adaptive hypotheses frequently invoked in an attempt to interpret the enameloid microstructure.

One of the authors of this book (SE) has sampled numerous living and fossil myliobatiform genera, including more than half of the species of Mobulidae (planktivorous forms) as well as several fossil forms, whose dentition shows a morphological transition between durophagous and planktivorous (e.g. *Burnhamia* and *Plinthicus*) (Enault, Adnet and Cappetta, 2013). At the time of their publication, these results confirmed several theories formulated in the recent literature. In particular, these observations indicated that a well-differentiated enameloid associated with the presence of PBE was specific to selachimorphs rather than to all neoselachians and that, with the exception of planktivorous forms, batomorph enameloid showed an ancestral morphology. The dental microstructures of the durophagous Myliobatiformes and Rajiformes sampled in this study were thus described as being composed of two distinct "layers", the homology of which to the structures known to exist in sharks (the "triple-layered enameloid") not being clearly established. Conversely, planktivorous Myliobatiformes systematically show a very thin, non-differentiated enameloid layer (SCE) (Figure 4.7). This apparently very simplified enameloid had been interpreted as a derived state resulting from the loss of one of the layers described in other batomorphs, a simplification considered to be related to the disappearance of the mechanical stresses imposed by a durophagous diet throughout the evolution of the group. In the genus *Manta*, one of the most derived planktivorous taxa, rows of the upper teeth were even lost secondarily as the group evolved (Adnet *et al.*, 2012; Marshall, 2009). The absence of functional stresses imposed by a "hard" diet may thus have led to the evolution of atypical morphologies and ornamentation patterns in certain planktivorous forms, and the principal conclusion of this study was that the enameloid evolution is more reflective of the ecological specializations of these animals than of their common evolutionary history.

However, some of these observations were proved to be inaccurate by subsequent discoveries, as demonstrated in an article co-written by the authors of this book (Enault *et al.*, 2015). These errors, due mainly to the incorrect interpretation of some images taken by an electron microscope, emphasize the difficulty of viewing certain histological features of a tissue such as enameloid. New data, as well as the re-examination of certain taxa, eventually revealed that all Myliobatiformes actually possess a thin, non-differentiated enameloid, as initially described in Mobulidae. It is always composed of a single unit with varying thickness, but relatively thinner than that described in other rays. This arrangement is preserved in all the living and fossil Myliobatiformes studied so far, regardless of their diet and dental morphology, which perfectly illustrates the fact that adaptive hypotheses centered exclusively on diet are not relevant.

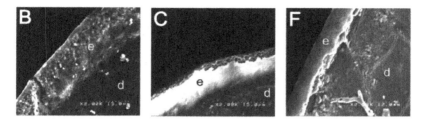

Figure 4.7. *Microstructure (cross-section) of the enameloid of three Mobulidae. Left to right:* Mobula japanica, Manta birostris *and* Archaeomanta sp., *an Eocene taxon. All these taxa show a thin layer of SCE. The sections have been treated for 5–10 s with 10% dilute HCl*

If a reduced, non-structured enameloid seems to be congruent with a diet that is not mechanically demanding, such as planktivory, the existence of this type of morphology in durophagous taxa is extremely surprising. Durophagy is indeed probably the diet imposing the greatest functional stresses, not only on dentition, but on the whole of the masticatory system, which includes the soft tissues associated with it. This type of specialized diet appears independently several times in batomorphs, particularly in the Rajiformes (*Rhina, Zapteryx*) and the Myliobatiformes (e.g. *Pastinachus*, Myliobatidae and Rhinopteridae), with durophagy probably having been inherited from a common ancestor in the two latter families (Dean *et al.*, 2007; Notarbartolo Di Sciara, 1987; Summers, 2000). Morphological adaptations to this type of diet in vertebrates include, in particular, a reduction in the number of teeth (which is highly visible in durophagous Myliobatiformes, especially *Aetobatus,* in which only the central tooth file remains) and underdeveloped cusps (Kolmann *et al.*, 2014). In durophagous Myliobatiformes, the functional stresses associated with this type of diet are compensated for by a thick tubular dentine, which probably allows the diffusion of compressive forces in the tissue. Other adaptations involve the masticatory system as a whole, shown notably by the fusion of symphyses and a very high degree of calcification in jaw cartilage (Dean *et al.*, 2007; Dingerkus *et al.*, 1991; Summers, 2000). The transition zone between the two main dental tissues, enameloid and dentin, which has been rarely studied and is generally poorly defined, may also provide new answers to the question of mechanical stress management during feeding. These adaptations help to compensate for highly reduced enameloid, the microstructure of which does not seem to be functionally beneficial in the context of a diet as mechanically demanding as durophagy. In addition, the oldest functional teeth are virtually devoid of enameloid because of their rapid abrasion when the animal feeds.

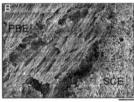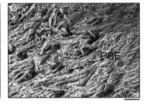

Figure 4.8. *Dental microstructure of planktivorous selachimorphs. A: Cetorhinus occidentalis, longitudinal cross-section. The scale bar is 10 μm. B: PBE and SCE of a specimen of* Rhincodon *sp. from the Belgian Miocene. The scale bar is 5 μm. C: PBE of* Rhincodon typus. *The scale bar is 6 μm*

The structure of the enameloid layer is particularly stable in Myliobatiformes, and is thus more reflective of their common evolutionary history than of the functional stresses associated with their specific diets. Similar observations have also been made in sharks (Figure 4.8), in which highly specialized diets have repeatedly appeared with some durophagous forms in the Heterodontiformes (*Heterodontus* spp.) and the Carcharhiniformes (*Mustelus* spp., *Sphyrna tiburo*), whereas planktivorous forms exist among the Orectolobiformes (*Rhincodon typus*) and Lamniformes (*Cetorhinus maximus, Megachasma pelagios*) (Cappetta, 2012). In all these genera (only *Megachasma* has not been studied, due to the rarity of the dental material), the characteristic structuring of selachimorph enameloid discussed in the previous chapters has been observed, and is thus retained regardless of the diet.

Although it is difficult to draw conclusions for all batomorphs as their enameloid microstructure presents a broad spectrum of variations, the example of the Myliobatiformes effectively illustrates the negligible impact of diet on this microstructure in this specific case and their extreme stability in terms of dental histology. Given the highly stable nature of the enameloid microstructure among selachimorphs as a whole and in the Myliobatiformes among rays, the use of the neoselachian enameloid microstructure in a taxonomic and/or phylogenetic context appears extremely promising for the future as long as the limitations addressed in this chapter can be effectively identified.

Enameloid Microstructure Diversity in Modern Shark Teeth

Modern sharks (selachimorphs) are the sister group of the batomorphs, both forming the neoselachian clade (modern selachians). With around 500 species, modern sharks represent only a small percentage of the wide variety of "fish" (chondrichthyans and osteichthyans), but this group displays a great diversity in terms of morphological and ecological adaptations as well as reproductive strategies. Though the vast majority of them are found in the marine realm, sharks are present in all aquatic environments, from freshwater to the deepest ocean depths (Ebert *et al.*, 2013). The first proven selachimorph fossils date from the early Permian (296 mya) of Russia, with teeth belonging to Synechodontiformes (Ivanov, 2005), a group of sharks with uncertain affinities that vanished in the early Cenozoic (60 mya). Despite their ancient origins, it was not until the early Jurassic (201 mya) that the selachimorphs began to diversify and move toward the top of the oceanic food chain. The majority of selachimorph groups are high-level trophic predators, ranging from durophagous sharks to ambush predators and mega-predators. However, some species (basking sharks, whale sharks and megamouth sharks) have developed microphagous (filter-feeding) diets. A wide range of predation strategies found in sharks are also illustrated by their main predation tool: their teeth. No less than nine types of dentition have been defined to illustrate the different adaptive patterns seen in shark teeth, taking into account the morphology, arrangement and heterodonty of the various teeth of the lower and upper jaws (Cappetta, 2012). These diverse morpho-adaptive dental patterns make sharks a subject of study particularly helpful in understanding the evolution of enameloid microstructure in relation to adaptive and/or ecological stresses.

5.1. Diversity and evolution of structures

Modern sharks are often considered to show the greatest diversity of enameloid structures among chondrichthyans. As we will see, the complexity of these structures and their organization is in fact significant in selachimorphs, but Chapter 4 of this book has shown that this is applicable to batomorphs as well. However, it is certain that the sampling of histological studies conducted to date has focused mainly on the selachimorphs. Therefore, the impression that the diversity of their enameloid is greater than that of other groups may be due in part to the overrepresentation of histological studies based on representatives of this group (compared with batomorphs in particular).

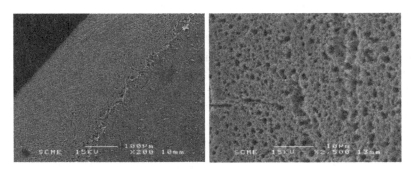

Figure 5.1. *Enameloid thickness in longitudinal cross-section of* Cretorectolobus *sp. from the Maastrichtian of Texas (left) and SCE in surface etching of* Synechodus *sp. from the Santonian of Kazakhstan (right)*

Since the work of Wolf-Ernst Reif in the 1970s, sharks' enameloid microstructure has been divided into three "layers" (Figure 5.1): an external layer (called SLE) and the two interior components of the BCE (PBE and TBE). The external layer is a typical SCE similar to the SCE of other chondrichthyans, made up of short, individualized, randomly oriented crystals (Figure 5.1). Its thickness varies according to taxa (5 µm in *Welcommia bodeuri;* 10 µm in *Paraorthacodus jurensis;* 6.5 µm in *Pachyhexanchus pockrandti;* 7 µm in *Synechodus* sp.) and also varies within each tooth, growing thicker from the base toward the apex of the crown. This layer is often difficult to see in cross-section due to functional wear as well as a possible poor compaction of the structure, all the more since prior acid etching is always necessary in order to observe the microstructure of a tooth via electron microscopy, and the latter can easily destroy the SCE. In any case, it represents only a small part of the thickness of the enameloid compared with the BCE. This layer, formed of bundles of crystals, is generally well developed in selachimorphs. The PBE component is made up of bundles oriented parallel to each other as well as to the surface of the crown (Figure 5.2). On the other hand, in most selachimorphs, this orientation changes near cutting edges, where the bundles are arranged perpendicular to the axis of the latter (variations do exist; see below). The TBE component is represented by entangled, randomly oriented bundles of crystals. The components of the BCE are not independent; only the orientation and organization change from one component to another and it is sometimes possible to follow one bundle as it goes from one component to the other (Figure 5.2). This close relationship between PBE and TBE is also represented at the base of the crown, where these two components are often indistinguishable from one another. Complexification occurs higher up in the crown, making it useful to take transverse cross-sections from the upper half of the crown in order to correctly identify the structures.

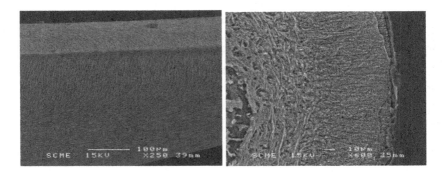

Figure 5.2. *Surface etching of a* Sphenodus nitidus *tooth from the Valanginian of France (left) showing parallel PBE bundles with apical–basal orientation, but oriented perpendicular to the cutting edge (top of image); and cross-section of a* Squatina sp. *tooth from the Miocene of France (right) showing continuity within the BCE between TBE bundles (left side of image) and PBE bundles (right side of image)*

The third and last component forming BCE in most selachimorphs is the RBE (*radial bundled enameloid*). This component combines multiple structures that have different characteristics but share the feature of passing through the enameloid perpendicular to the surface of the crown. Of these structures, radial microcrystals (LRB – *loose radial bundles*) start in the SCE and penetrate the bundles of the PBE in a perpendicular fashion (Figure 5.3). Though these structures frequently seem to be present in selachimorphs, they remain difficult to see due to the poor compaction of microcrystals between them, and these radial microcrystals are often better observed through surface etchings. In some taxa, these microcrystals are densely amalgamated into radial bundles (or TRB – *thick radial bundles*), which penetrate the PBE and can also penetrate the TBE (Figure 5.3). Whether in radial microcrystals or radial bundles, these structures probably originate in the SCE (Guinot and Cappetta 2011; Enault *et al.* 2015). This raises some questions at the developmental level, since the SCE layer seems to have its origins in the ectoderm as the protein matrix is secreted by ameloblasts, while the BCE may be mesodermal in origin with a protein matrix secreted by odontoblasts. Thus, we would have crystals generated by an ectodermal matrix included in other crystals generated by a mesodermal matrix. Given the centripetal mineralization of enameloid, this would imply that the protein matrices of SCE and BCE are secreted simultaneously. However, the most surprising observation concerns microcrystals forming radial bundles which, despite their origin in the SCE, are organized in bundles, which is normally a characteristic of the BCE layer. One possible explanation would be that even though the microcrystals that form radial bundles are created by a protein matrix secreted by ameloblasts, their arrangement in bundles would be caused by a signal given by the protein matrix of mesodermal origin, before mineralization. The BCE would thus be a zone in which the mineralization of microcrystals, whether generated by matrices of ectodermal or mesodermal origin, would occur in bundle-type formation.

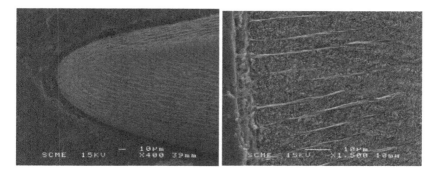

Figure 5.3. *Surface etching of a* Welcommia bodeuri *tooth from the Valanginian of France (left) showing radial microcrystals perpendicular to the PBE bundles; and transverse cross-section of a* Synechodus sp. *tooth from the Albian of Kazakhstan (right) showing radial bundles (TBE) originating from the SCE and crossing the PBE in a perpendicular fashion*

Other, less typical radial structures are known to exist in some selachimorphs. These include amalgamated crystals (Figure 5.4), which have been described in Cretaceous sharks including *Welcommia bodeuri* and *Paraorthacodus jurensis*

(Guinot and Cappetta 2011). These structures are composed of wide, flat amalgamated crystals (around 4 µm wide) that cross the PBE bundles in a perpendicular manner and tilt toward the base of the crown, penetrating the TBE. It appears that these amalgamated crystals are confined to the BCE, with no connection to the SCE. In *Pachyhexanchus pockrandti*, a Cretaceous hexanchiform, these crystals are less fused and the flat structures they form are shorter but wider (around 15 µm). Radial microcrystals, radial bundles and amalgamated crystals all originate in or around the SCE and penetrate the PBE in a perpendicular manner to reach the TBE. It is therefore probable that, through this intertwining of bundles/crystals, these structures help to bind the enameloid layer together as well as contributing greater resistance to the tensile forces applied to the crown during predation. On the other hand, certain taxa seem to show a tendency toward the reduction of their enameloid via the presence of radial grooves crossing the PBE and/or holes at the SCE/PBE boundary (Figure 5.4). It is more plausible to consider these holes and radial grooves as the imprints of structures formed by organic elements which subsequently dissolved.

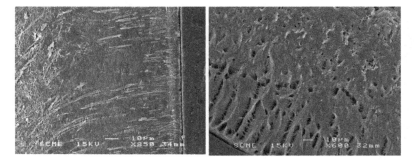

Figure 5.4. *Longitudinal cross-section of a* Welcommia bodeuri *tooth from the Valanginian of France showing amalgamated crystals passing through the PBE and tilting toward the base of the tooth as they get closer to the TBE (left); and transverse section of a tooth from the same species (right) showing grooves in the PBE and holes under the SCE/PBE boundary*

Shark teeth frequently have sharp cutting edges (mesial and distal) as well as ornamentation ridges and other carving features on the lingual and/or labial surfaces. We have seen that the cutting edges are generally formed by PBE bundles that change orientation to lie perpendicular to the axis of the cutting edge and are covered by SCE with uniform thickness. However, variations in the organization of cutting edge enameloid have been observed, despite the small amount of sectioning conducted. In *Rhomphaiodon*, for example, a synechodontiform from the late Triassic, it has been shown that the cutting edges are formed by PBE bundles (oriented perpendicular to the axis of the cutting edge and parallel to the surface of the crown) and also by a thickening of the SCE. Similarly, ornamentation ridges are commonly formed by a thickening of the SCE in selachimorphs. However, in certain taxa such as *Striatolamia macrota*, ornamentation ridges seem to be formed solely by the puckering of PBE bundles oriented in an apical–basal manner, with the SCE retaining a constant thickness. The positioning of cutting edges and ridges seems, then, to result from two different mechanisms: a puckering or thickening of the enameloid, whether the external layer (SCE) or the internal one (PBE), to form ridges, and a change in the orientation of PBE bundles, combined (or not) with a thickening of the SCE for cutting edges. However, this model is contradicted by studies (Cuny *et al.* 1998; Cuny and Risnes 2005), which have demonstrated that the microstructure of the ornamentation ridges in certain taxa (*Rhomphaiodon minor, Grozonodon candaui*) is composed of PBE bundles perpendicular to their axis and associated with a thickening of the SCE, exactly like cutting edges. The same may be true for other taxa (*Hueneichthys costatus*), but these observations have been made by surface studies only, which complicates our understanding of the structures. To make everything even more complex, the mesial cutting edges of *Sphenodus nitidus* present a classic structure with PBE bundles perpendicular to the axis of the cutting edge and an

SLE of constant thickness. On the other hand, the more internal PBE bundles show no change in orientation, but only a simple thickening. The term RCEL (for *Ridges/ Cutting Edge Layer*) had been coined to define the structures associated with cutting edges and ornamentation ridges (Cuny and Risnes 2005), since these structures seemed quite distinct from the BCE. However, this clean-seeming separation was almost certainly the result of observation artifacts having to do with the change in orientation of PBE bundles, and it is more likely that the RCEL is simply a joining of the SCE and BCE layers, and not just made of the SCE. Clearly, the diversity of cutting edge and ornamentation ridge structures is not yet fully understood, and neither is their distribution within the various groups of chondrichthyans. Yet, as Cuny and Risnes (2005) have emphasized, these differences have consequences in terms of the details of tooth formation, since enameloid mineralization seems to begin in the cutting edges. Ridges with cutting edge-type microstructures imply that there are as many mineralization initiation sites as there are ornamentation ridges. Conversely, a simple puckering of the enameloid surface would have no consequences in terms of changes in the number of sites of mineralization initiation. Thus, the appearance of cutting edges and ornamentation ridges may have taken place multiple times independently of one another in chondrichthyans.

The diversity and complexity of the microstructures we have just seen are found mainly in derived selachimorphs, but what of the more basal selachimorphs? The problems encountered while attempting to answer this question result from the difficulty of identifying the systematic position of these basal neoselachians. For example, the species often considered to be the oldest among the selachimorphs, *Synechodus antiquus* from the early Permian of Russia (Ivanov 2005), belongs to the order Synechodontiformes. However, the phylogenetic position of this group remains disputed: whether galeomorph or squalomorph or stem selachimorph. Thus, it is difficult to draw conclusions about the ancestral state of selachimorph enameloid on the basis of these taxa. We would also add that the genus *Synechodus* has a huge stratigraphic distribution (around 230 myrs) and, despite the small amount of sampling for this genus, a wide variety of microstructures have been shown to exist (compare the enameloid of *Synechodus rhaeticus* described by Cuny and Risnes (2005) and that of *Synechodus* sp. described by Guinot and Cappetta (2011)). There are some other known taxa in the fossil record that may represent basal selachimorphs; the cases of *Mcmurdodus* (Devonian), *Cooleyella* and *Ginteria* (Anachronistidae of the Carboniferous) were discussed in Chapter 2 of this book. The Anachronistidae, the enameloid of which is composed of poorly defined bundles perpendicular to the surface, are generally considered to be neoselachians with uncertain affinities (possibly batomorphs) based on the characteristics of their tooth roots, particularly their vascularization. The European Triassic has yielded several dental remains of taxa with uncertain affinities, but which are generally ranked among the (stem) neoselachians. Among these, we have briefly mentioned the microstructure of the cutting tooth edges of the species *Hueneichthys costatus*, known from the Rhaetian via a single, rootless tooth (Reif 1977). This species also possesses an external SCE and a PBE with radial bundles, but there is no mention of the TBE component, which may have to do with the fact that this surface study did not explore the deeper tissues. The teeth of the genera *Doratodus* and *Vallisia* have neoselachian morphologies, but an enameloid composed solely of SCE (Duffin 1981; Cuny and Benton 1999), the outermost part of which contains crystals oriented perpendicular to the enameloid surface in *Vallisia*. Nevertheless, considerations of the enameloid of these two taxa must be treated cautiously, as their microstructure has only been explored via surface studies (classic or fracture surface). *Reifia minuta,* a species from the late Triassic of Germany, shows crown and root morphology similar to what can be seen in certain selachimorphs. One histological study (Duffin 1980) has demonstrated the presence of a classic PBE beneath a thin layer of SCE, the presence of which has not been proven. More surprising still, this study indicates the presence of an internal layer composed of individualized, randomly oriented crystals beneath the PBE. Here again, only a surface study of a single tooth has been conducted, and this specific microstructure with SCE-type tissue beneath the PBE has yet to be confirmed. The teeth of the species *Pseudocetorhinus pickfordi* (Rhaetian), initially placed in the family of modern basking sharks (Cetorhinidae), and later considered to be a neoselachian with uncertain affinities, have been the subject of numerous histological studies (Cuny 1998; Duffin 1998; Cuny and Benton 1999; Cuny *et al.* 2000). Though all of these studies are based on surface explorations, their large number has given us a good understanding of the enameloid structure of this taxon, which is composed of classic neoselachian layers and components (SCE, PBE, TBE and radial bundles). The crystals of the SCE, however, show a preferential orientation perpendicular to the surface of the crown. It should be noted that the teeth of *P. pickfordi* are characterized by a high degree of heterodonty, with high, sharp front teeth and grinding-type lateral teeth, which are low and extended in a mesial–distal manner. However, microstructural studies do not indicate the position of the teeth studied, which could be important to the structures observed. Another contemporaneous species, *Duffinselache holwellensis* (probably similar to, if not synonymous with, *P. pickfordi*), shows a microstructure similar to *P. pickfordi*, except that the radial bundles seem to be absent in *D. holwellensis*. Finally, *Pseudodalatias*, known from the two species *P. barnstonensis* (Norian–Rhaetian) and *P. henarejensis* (Ladinian), is also a genus with uncertain affinities. It was originally classified as part of the modern genus *Dalatias* (kitefin sharks) due to its vaguely similar dental morphology, and was then grouped with the hybodonts before this placement became doubtful as well. Unlike the Triassic taxa examined previously, the teeth of *Pseudodalatias* show a high degree of specialization toward grasping–cutting type dentition with teeth compressed labial-lingually and serrations on some teeth (possibly belonging to the lower jaws). The enameloid of this taxon shows an astonishing microstructure composed of an SCE in which the outermost crystals are parallel to the surface of the crown but are oriented perpendicular to the surface in the internal part of the enameloid.

Despite the low taxonomic diversity of the earliest neoselachians, we have just seen that they represent a wide range of microstructures in their dental enameloid. On the other hand, except for *Pseudocetorhinus* and possibly *Hueneichthys*, all of these taxa show a simpler microstructure than that of the derived selachimorphs. Unfortunately, the phylogenetic relationships of these taxa with the major groups of modern selachians are difficult to establish on the basis of their dental morphology alone, which hinders our understanding of the distribution and polarity of the structures at the base of the selachian phylogenetic tree. In addition, these studies of dental microstructure were originally undertaken with the objective of placing these taxa more precisely within the selachians, before it was realized that the characteristics utilized were not sufficient for this purpose. Moreover, most of the histological studies focusing on these taxa were based only on surface studies, though, as we have seen in Chapter 3 of this book, certain structures, such as organizations of loosely bound bundles, cannot be detected during this type of exploration. The lack of resolution with regard to the relationships among fossil chondrichthyans, added to the small amount of sampling from these various groups as well as the problems with identification of structures encountered when using certain methods (surface exploration), means that, for the moment, it is impossible for us to suggest a solid scenario explaining the polarity of histological characteristics in chondrichthyans. Thus, even though the Carboniferous–Triassic period seems to have been a major period of evolution for chondrichthyans, it remains a gray area in terms of our understanding of the evolution of their enameloid microstructure. However, in the light of what we have been able to show in this book, it is clear that hybodonts and batomorphs show a more differentiated microstructure than what was thought previously, making them more similar to what we have just seen for selachimorphs. The presence in hybodonts and batomorphs of loose bundles of crystals perpendicular to the surface of the crown (identified here as RBE) seems to represent a plesiomorphic characteristic, especially because they are present in the dermal denticles of certain hybodonts. However, these bundles tend to be oriented parallel to the surface of the crown in some batomorphs, particularly at the edges of longitudinal crests (see Chapter 4), recalling the PBE of selachimorphs. Could it be that these bundles perpendicular to the surface are a PBE in which the bundles do not converge toward the apex due to the morphology of the teeth of these groups, which are in general low and possess only a longitudinal crest? Whatever the case, the presence of PBE-type bundles in selachimorphs, in some cladodontomorphs, and possibly in some batomorphs, remains difficult to interpret for the time being, taking into account the lack of a clear understanding of the relationships within the cladodontomorphs and between the cladodontomorphs and the neoselachians. In light of this observation, it is surely necessary to increase sampling within the various groups of fossil and living chondrichthyans in order to identify enameloid characteristics that can define and distinguish between neoselachians, selachimorphs and batomorphs.

5.2. Serrated teeth and mega-predation

Modern sharks are distinguished from the other groups of chondrichthyans by a wide variety of their methods of predation, represented by the many types of teeth they produce. Among these, cutting-type teeth are associated with a carnivorous diet, which corresponds to active predators or scavengers. This type of dentition is formed of high teeth with crowns that are mesio-distally elongated and labio-lingually compressed, though generally curved toward the commissure. Cutting-type teeth are frequently seen in modern (great white and mako sharks) and fossil (*Squalicorax*) Lamniformes, as well as in Carcharhiniformes (tiger sharks), Squaliformes (*Squalus, Cirrhigaleus*) and Echinorhiniformes (bramble sharks). There is a variant of this dental type in which the teeth of only one jaw (lower or upper depending on species) possess cutting-type morphology, while the teeth of the other jaw are narrower and are used to clutch prey. Similar to the *sensu stricto* cutting type of dentition, the cutting–clutching dentition is present in a large variety of unrelated groups, including some Carcharhiniformes (Carcharhinidae, Hemigaleidae, Sphyrnidae) and certain Squaliformes and Hexanchiformes. In both of these types of teeth, the cutting edges of cutting teeth frequently have serrations which, like those of a steak knife, help to cut up prey. These cutting-type morphologies appeared multiple times independently of one another throughout the evolution of different groups of selachimorphs. Within these different "lineages", the basal species show cutting teeth with smooth cutting edges, whereas serrations appear in the more derived species and thus after the compaction of teeth. Though cutting teeth are widespread in selachimorphs, as we saw in Chapter 3 of this book, this is not the case with other chondrichthyans. Only the genera *Carcharopsis* among the Palaeozoic sharks, *Priohybodus* among the hybodonts and *Pseudodalatias* developed this type of tooth. There are other hybodonts with cutting dentition (*Pororhiza, Mukdahanodus, Thaiodus*), but they retained a low crown. How can we explain the rarity of this dental type in the other groups of chondrichthyans, while it appeared several times in selachimorphs and corresponds to an enormous range of species in this group? Are these serrations all identical in the different taxa that developed this type of tooth?

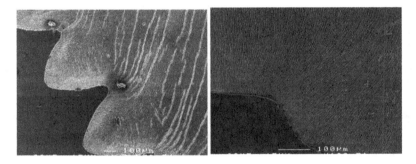

Figure 5.5. *Arrangement of PBE bundles in the serrations of a modern great white shark tooth (*Carcharodon carcharias)* showing primary and secondary changes in orientation*

The question of the enameloid microstructure of tooth serrations in selachimorphs was first addressed by Andreev (2010). In this study, the author focused on the structure of the cutting edges of the teeth of unrelated groups: a fossil lamniform from the Cretaceous (*Squalicorax*) as well as three fossil carcharhiniform species from the Miocene (*Hemipristis serra, Galeocerdo* sp. and *Carcharhinus* sp.). As with the teeth of species that lack serrations, the cutting edges of these four taxa are formed by PBE bundles which change orientation at the outermost parts of the cutting edges to lie perpendicular to the long axis of the tooth. This classic change of orientation in selachimorphs is referred to as a primary change. In *Squalicorax*, the serrations consist simply of indentations in the PBE and, in both the concavities and convexities of the serrations, the PBE bundles show only a primary change. In addition to the primary change, the serrations of *Galeocerdo* and *Carcharhinus* feature a secondary direction change of bundles that occurs as soon as these bundles penetrate the serrated area. This secondary change results in a perpendicular orientation of the bundles in relation to the edges of each serration. The bundles of *Hemipristis* show an organization similar to what has been described for *Squalicorax* but only in the upper part of the tooth, where the serrations are less marked. The deep serrations in the lower parts show a secondary change (in the serration), but one that occurs closer to the edge than in *Galeocerdo* and *Carcharhinus*. Other examples of double orientation changes in PBE bundles that form serrations were reported by Reif (1973) in other taxa such as *Oxynotus* and *Dalatias* (Squaliformes) as well as in the lamniform *Carcharodon carcharias* (Figure 5.5). Thus, a degree of variability exists in the microstructure of serrations among selachimorphs, and the reason for this variability may be explained by tooth size. The *Squalicorax* teeth studies showing a primary change are small, with shallow serrations. Similarly, serrations with a primary change in the apical part of *Hemipristis* teeth are shallow, whereas the deeper ones in the basal part show two changes in orientation. The effect of size could be confirmed by the presence of double orientation changes in the PBE of the serrations of a larger species of *Squalicorax* (Reif 1973), but no images have been published. In any case, the many independent appearances of serrations observed in the evolutionary history of selachimorphs systematically involve PBE, and in only two different arrangements.

The role of PBE in the formation of serrations may help explain the frequency of the production of this type of cutting tooth in selachimorphs compared with other groups of chondrichthyans. However, we have seen in this book that certain chondrichthyans whose tooth enameloid is devoid of PBE have produced this type of tooth. In these taxa (*Carcharopsis, Priohybodus, Pseudodalatias*), the enameloid is composed of an inner layer of crystals perpendicular to the tooth surface, beneath an outer layer of SCE. It seems, then, that the presence of PBE is not a prerequisite for labio–lingual flattening, or for the development of serrations. The reasons for the rarity of cutting-type teeth in non-selachimorph chondrichthyans are thus not linked solely to the condition of their enameloid, though these adaptations necessitate a significant change in crystal organization. Selachimorphs showing cutting-type (or related) dental morphologies have developed a carnivorous diet represented by both active predators and opportunistic predators/scavengers. The general morphology of hybodonts and other Palaeozoic chondrichthyans indicates that these groups were not able to swim as actively as is seen in certain living Lamniformes and Carcharhiniformes. The rare examples of cutting-type teeth in non-selachimorph chondrichthyans would be more likely to correspond with scavengers. As noted in Chapter 3, fossil remains of cutting-type teeth from hybodonts have been restricted to continental aquatic environments and stratigraphically confined to the period between the late Jurassic and the middle Cretaceous. The development of this type of highly specialized predation in continental aquatic environments subject to intense adaptative pressures and competition may explain the rarity of this type of dentition. On the other hand, the reason why marine hybodonts and Palaeozoic chondrichthyans did not develop, or rarely developed (*Carcharopsis*) this type of dentition remains a mystery. However, we must not overlook the Eugeneodontiformes, which, despite the arrangement of their teeth in spirals resembling a circular saw, developed high, flat, and serrated teeth. The microstructure of the tooth enameloid of Eugeneodontiformes remains enigmatic, and the rare studies that have been carried out are difficult to interpret. The rarity of cutting-type teeth in the marine environment in pre-Cretaceous chondrichthyans may have more to do with ecological and competitive stresses rather than adaptive ones.

5.3. Example of adaptation to durophagy: bullhead sharks

Bullhead sharks (Heterodontiformes) are a group of modern selachimorphs that show little diversity (nine species) and are confined to tropical and intertropical areas. They are benthic predators whose food is composed mainly of invertebrates possessing hard exoskeletons (sea urchins, bivalves, various crustaceans, gastropods). These sharks are unique due to their dentition, which is composed of front clutching teeth (compressed labio-lingually and with one or more cusps) and lateral molariform grinding-type teeth (flattened apico-basally and elongated anteroposteriorly). This difference in dental morphology within a single jaw, called monognathic heterodonty, is the source of the name of the bullhead shark: *Heterodontus* (from the Greek for "different teeth"). *Heterodontus* is the only living representative of the order Heterodontiformes. However, this monognathic heterodonty is seen only in adult individuals, with the entire dentition of juveniles composed of clutching teeth, though these feature a larger number of cuspids than those of adults. During ontogenesis, these lateral files of clutching teeth gradually change, becoming lower and more elongated until they eventually achieve a molariform morphology. In addition to the files of teeth present in juveniles and retained in the adult state, additional files (anterior and lateral) appear during ontogenesis.

These morphological differences within one jaw of a single individual offer a unique opportunity to explore intra-individual variations in enameloid microstructure in chondrichthyans. Reif (1973) was the first to show the histological differences between anterior and lateral teeth. The anterior teeth of *Heterodontus* have a classic selachimorph microstructure (SCE, PBE, TBE) as well as a RBE component (Figure 5.6). Conversely, the lateral teeth consist only of a thick TBE covered by an SCE; the PBE is absent. The absence of PBE in these lateral grinding teeth has been interpreted as an adaptation to durophagy, with this component probably being a structure helping to resist tensile forces, while TBE would be more resistant to compressive stresses (Preuschoft *et al.* 1974). This example shows the plasticity of enameloid structures, since the teeth of a single jaw show different microstructures (loss of a component) depending on their morphology, and thus their function. This is even more striking because this differentiation occurs during ontogenesis on files of teeth that include anterior teeth, the microstructure of which should thus include three layers (though the anterior teeth of juvenile *Heterodontus* have not been examined to date).

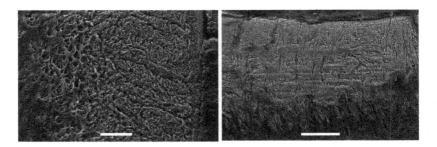

Figure 5.6. *Difference in enameloid microstructure between an anterior tooth (left, transverse section, scale = 20 μm) and a lateral tooth (right, axial labial–lingual section, scale = 50 μm) of* Heterodontus francisci, *a modern bullhead shark*

Other durophagous groups show a microstructure comparable to that of the lateral teeth of *Heterodontus*. Among hybodonts, we have seen the examples of *Heteroptychodus* and *Asteracanthus*, the teeth of which are composed of an SCE and a TBE (in some of these taxa, well-developed dentine tubules are also present). Recently published results (Hoffman *et al.* 2016) on *Ptychodus* teeth with a high crown with PBE between the SCE and TBE may represent a case similar to that of *Heterodontus*, and it would be relevant to explore the microstructure of very lateral teeth (with a lower crown) in the same species as the one studied by Professor Hoffman's group. More generally, there are a large number of modern and fossil selachimorph groups showing heterodonty that includes low lateral/posterior teeth lacking cusps, such as *Ptychocorax*, *Synechodus* (*S. dubrisiensis*) and *Sphyrna* (*S. tiburo*), which would be helpful to study. In any case, the data obtained for hybodonts and *Heterodontus* indicate that the absence of parallel bundles and the sole presence of SCE+TBE are indeed the result of adaptation to a durophagous diet. This confirms the results obtained by Preuschoft *et al.* (1974), who suggested a tensile force resistance function for PBE and a compression force resistance function for TBE, with SCE limiting the propagation of fissures.

However, in the last chapter, we saw the lack of variations in the enameloid microstructure of certain groups of batomorphs, such as the Myliobatiformes, the enameloid of which is composed of a single SCE, whatever the diet of the taxa. The absence of TBE in durophagous Myliobatiformes might be explained by the fact that the BCE layer has been lost before the specialization of some representatives of this group toward durophagy. However, these groups then became specialized by varying other characteristics such as the thickness and structure of dentine or the interlocking of teeth (see Chapter 4) to compensate for compressive stresses induced during predation.

Comparison of Enameloid Microstructure in Actinopterygian and Elasmobranch Teeth

6.1. Ganoine and acrodine

Although there are a wide variety of microstructures (see previous chapters), the hypermineralized tissues of elasmobranchs follow the same general pattern. However, the situation is somewhat more complex within the actinopterygians (ray-finned bony fish), which exhibit two types of hypermineralized tissues, ganoin and acrodin (Francillon-Vieillot *et al.*, 1990), with the latter being considered a form of enameloid. Ganoin covers certain types of scales and the lower part of teeth. In 2015, the work of a Swedish team from Uppsala University showed the presence of genes in the skin of the spotted gar (*Lepisosteus oculatus)* coding two of the proteins (EMP or enamel matrix proteins) necessary for the formation of the tooth enamel in tetrapods, enamelin and ameloblastin (Qu *et al.*, 2015). This discovery strengthens the argument for the homology between ganoin and enamel, which was previously suggested by other studies (Sire *et al.* 1987; Sire, 1995; Zylberberg *et al.* 1997; Sire, Donoghue and Vickaryous 2009). By extension, the same is thought to be true for ganoin present in teeth, although not every author is in agreement with this point (Donoghue *et al.*, 2006). In any case, unlike enamel, the microcrystals that make up ganoin are not arranged in prisms, but show a stratified structure (Figure 6.1).

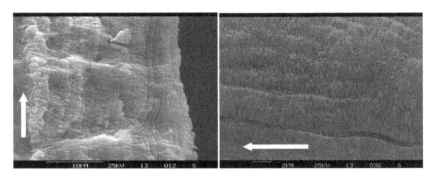

Figure 6.1. *Stratified microstructure of ganoin in two fish from the late Triassic of Europe. Left: cross-section of ganoin on an incisiform tooth of* Sargodon tomicus *from the Upper Triassic (Rhaetian) of Aust Cliff, England, seen in a tooth fracture treated for 5 s with 10% dilute HCl. Right: cross-section of ganoin on a* Severnichthys acuminatus *tooth from the Rhaetian of Saint-Nicolas-de-Port, France, treated for 35 s with 10% dilute HCl. The arrow indicates the direction of the tooth apex*

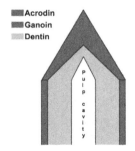

Figure 6.2. *Schematic longitudinal cross-section of an actinopterygian tooth showing the distribution of the various tissues that it contains. For a color version of this figure, see www.iste.co.uk/cuny/selachians.zip*

In the fossil record, ganoin first appears on scales and later on teeth. *Andreolepis*, a bony fish from the Silurian of Russia and Northern Europe, actually shows scales covered with ganoine, while its teeth appear to lack any hypermineralized tissue and are composed solely of dentin (Chen *et al.*, 2016). Over the course of their evolution, ganoin tends to disappear from both teeth and scales in teleosteans, which are derived actinopterygians that represent more than 99% of living species of fish (trout, tuna and cod are teleosteans) (Qu *et al.*, 2015; Schultze, 2016).

The second type of hypermineralized tissue, acrodin, forms a cap covering the apical part of the tooth (Figure 6.2). Unlike ganoin, acrodin is not stratified but is formed of bundles of microcrystals, like the enameloid of elasmobranchs (Andreev, 2011). Acrodin is thus often considered to be an enameloid.

Acrodin, like elasmobranch enameloid, begins to mineralize in a centripetal manner before the dentin, while tissues that are purely epithelial in origin, like enamel and ganoin, mineralize centrifugally only after the latter. These similarities in the structure and development between shark enameloid and actinopterygian acrodin suggest that these two tissues share a common evolutionary origin. This hypothesis is particularly supported by the fact that enameloid-type tissues, which result from interactions between ameloblasts and odontoblasts, have been shown to exist in vertebrates before the appearance of jaws, such as in *Astraspis*, an agnathan from the Devonian of the Americas (Sansom *et al.*, 1997). However, acrodin seems to be absent in the teeth of ancient actinopterygians, such as *Cheirolepis*, from the Devonian of Europe and North America (Gardiner *et al.*, 2005), while the teeth of certain forms of selachians also seem to lack enameloid, which we addressed in a previous chapter. An observation such as this tends to show that the appearance of acrodin actually takes place after that of ganoin in the evolution of actinopterygians. This late appearance makes it difficult to believe that it represents a common inheritance that appeared before the separation between bony and cartilaginous fish. Moreover, other differences have been noted between acrodin and the enameloid of elasmobranchs. These include their composition: the former is composed of hydroxyapatite microcrystals, whereas the latter is made of fluorapatite. Their protein matrix also appears different, with collagen more abundant in the acrodin matrix (Berkovitz and Shellis, 2017). Thus, the homology between acrodin and elasmobranch enameloid remains difficult to demonstrate given our current understanding of their evolution and composition.

6.2. Comparison of acrodin and elasmobranch enameloid

Unlike the teeth of elasmobranchs, actinopterygian teeth seem to lack a surface layer of enameloid composed of simple SCE-type crystals. Acrodin thus belongs to the category of bundled enameloids (BCE). It is noteworthy, however, that a recent study by the Japanese team of Ichiro Sasagawa has indicated the presence of a thin layer of ganoin in *Polypterus senegalus* covering the acrodin cap, which may correspond to the SCE in elasmobranchs (Sasagawa *et al.*, 2013). This topology again raises the question of the homology of SCE with a tissue that is exclusively epithelial in origin.

The presence of bundles of microcrystals more or less entangled within the acrodin of fossil actinopterygian teeth, recalling the TBE of elasmobranchs, was noted in the early 1970s by the Swedish paleontologist Tor Ørvig (Ørvig, 1973). In the late 1970s, other researchers, including the German Wolf-Ernst Reif and Peter Shellis in the United Kingdom observed in the teeth of two families of teleosteans, the Characidae (the piranha family) and the Sphyraenidae (the barracuda family), the presence of bundles parallel to the surface and to the axis of the tooth, very similar to the PBE of certain ctenacanth sharks and selachimorphs (see previous chapters) (Shellis and Berkovitz, 1976; Reif, 1979). A PBE was also shown to exist in several species of Lophiiformes, the anglerfish (*Lophius piscatorius*), the blackbellied angler (*Lophius budegassa*) and the American angler (*Lophius americanus*), by a team of French researchers in 1980 (Kerebel and Cabellec, 1980). According to the work of Reif and Shellis, the acrodine appears better organized in its outer layer than in its inner one, and its overall structure is similar to that observed in selachimorphs, with one outer layer made of parallel bundles (PBE) (when present) and one inner layer made of tangled bundles (TBE). However, the radial component (RBE) of the PBE appears to be less developed in actinopterygians than in selachimorphs. When the PBE is absent, the TBE is better structured in its outer layer than in its inner layer.

The presence of PBE has been demonstrated, to date, only in conical teeth (barracuda; anglerfish) and teeth compressed labio-lingually (piranhas), but seems to be absent in flatter and more bulbous teeth, in which the acrodin is composed mainly of TBE (Andreev, 2011). Due to the entanglement of these bundles, this TBE provides a high degree of resistance to compressive forces, whereas PBE offers better resistance to tensile forces. Thus, their mechanical characteristics explain their distribution according to different dental morphologies. A very similar phenomenon is seen in some selachimorphs, with PBE disappearing in the posterior grinding teeth of bullhead sharks of genus *Heterodontus* (see Chapter 5), although this phenomenon does not occur in species with dentition made up partially or totally of crushing teeth (*Sphyrna tiburo*, species of the genus *Mustelus*, etc.). Crushing teeth are not accompanied by an increase in tooth size, unlike what we see in the grinding dentition of *Heterodontus*, which may explain the differences in microstructure between the two types of teeth. The lack of experimental data, particularly concerning an *in situ* study of the mechanical resistance of the various components of hypermineralized tissues, makes it difficult to demonstrate the correlations between morphology, tooth function and the microstructure of acrodine (and other types of enameloid) with certainty.

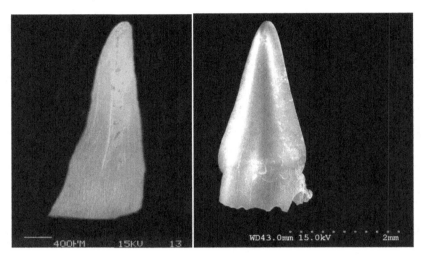

Figure 6.3. *Left:* Severnichthys acuminatus *tooth from the Rhaetian of Lons-le-Saunier, France, in mesial or distal view. Right:* Caturus *sp. tooth from the Upper Cretaceous of the region of Tataouine (Tunisia), in labial or lingual view*

Our team's research has confirmed the presence of PBE in the conical teeth we have studied; however, it has also showed significant variations in the organization of this PBE. For example, in teeth possessing well-developed cutting edges, such as the teeth of *Caturus* from the Lower Cretaceous of Tunisia (Figure 6.3), the bundles forming PBE show a clear shift in orientation at the location of these cutting edges and become perpendicular to the tooth axis (Figure 6.4) (Cuny *et al.*, 2010). Similar observations have been made concerning the cutting edges of selachimorph teeth (see Chapter 5) and it is possible that this change in orientation helps to reinforce the cutting edges mechanically. In some selachimorphs with highly ornamented teeth, the same phenomenon is seen at the location of ornamentation ridges (see Chapter 5). An unexpected variant of this phenomenon has been observed in teeth from the European Late Triassic belonging to the species *Severnichthys acuminatus*. This species shows conical teeth, in which the acrodine cap can show well-developed ornamentation ridges on its lingual surface (Figure 6.3). A change in the orientation of the PBE bundles also occurs in these ridges. However, comparing this to what has been observed in the ornamentation ridges of selachimorphs reveals two significant differences.

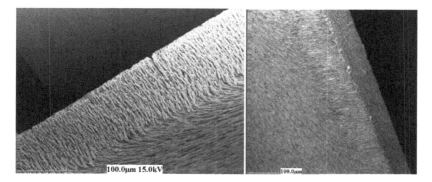

Figure 6.4. *Change in orientation of PBE bundles at the location of the cutting edge on a* Caturus *tooth from the Lower Cretaceous of Tunisia. Left: detail of the cutting edge on a tooth treated for 35 s in 10% dilute HCl. Right: same tooth treated for an additional minute in the same acid*

First, in selachimorphs, this change in PBE bundle orientation is confined strictly to the ridge itself. In *Severnichthys acuminatus,* the bundles forming the ridges meet between these ridges to form a distinct layer, in which all the bundles are oriented perpendicular to the tooth axis. Beneath this layer is a more classic PBE in which the bundles are oriented parallel to the axis (Figure 6.5). In this way, the PBE is separated into two superimposed components, the bundles of which are perpendicular to each other.

Second, in selachimorphs, the ornamentation ridges are covered by a thickened SCE (see Chapter 5). A component of this type has never been observed in *Severnichthys acuminatus,* which seems to confirm the absence of an SCE-type surface layer within the acrodin. It is difficult to explain these differences between selachimorphs and actinopterygians, which may be due to developmental stresses as much as to the results of selection related to the mechanical resistance of these different configurations. Given our current degree of understanding, it is extremely difficult to favor one or the other of these theories. In addition, although the microstructure of ridges is well understood in many elasmobranchs, this is not the case with the actinopterygians, among which only *Severnichthys* has been studied in detail. Thus, we do not know if this is a widely occurring structure or if it represents only one specific case. It will be difficult to make progress with regard to this question as long as the sampling problem remains unresolved.

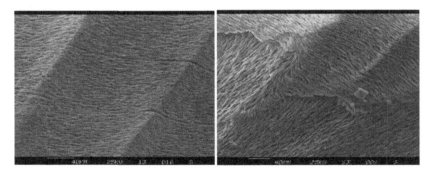

Figure 6.5. *Microstructure of ornamentation ridges in a* Severnichthys acuminatus *tooth from the Rhaetian of Aust Cliff, England. Left: PBE bundles showing orientation perpendicular to the tooth axis and joined from one ridge to the next. Tooth treated for 30 s in 10% dilute HCl. Right: PBE layer in which the bundles are oriented parallel to the tooth axis, appearing beneath the PBE layer forming ridges. Tooth treated for 1 min in 10% dilute HCl*

In some chondrichthyans, particularly hybodonts, the lower part of the enameloid may be penetrated by tubules originating in the dentin, which blurs the boundary between the enameloid and dentin (see Chapter 3). A similar phenomenon has been observed in some rare fossil actinopterygians. The acrodin in the teeth of *Sargodon tomicus* from the European Rhaetian shows well-defined canals in its lower part (Figure 6.6). This feature has been used since the late 19th Century to identify fossil teeth of this species found isolated (Andreev, 2011). Inside each of these canals are small tubules arranged radially, which probably housed odontoblastic extensions during the animal's lifetime. These canals probably contained blood vessels and odontoblasts, which further supports the theory that the latter contributed to acrodin production.

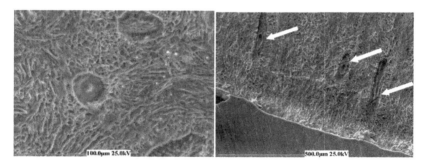

Figure 6.6. *Left: surface of a molariform tooth of* Sargodon tomicus *from the Rhaetian of Bristol, England, showing its canals in the midst of tangled acrodine bundles. Tooth treated for 35 s in 10% dilute HCl. Right: cross-section of a* Ptychodus *tooth of unknown origin. The white arrows indicate the placement of similar canals within its enameloid. The tooth surface is pointing toward the bottom of the image. Tooth treated for 30 s in 10% dilute HCl*

In 2011, the research of Plamen Andreev also showed similar canals in another group of actinopterygians possessing grinding teeth: pycnodonts (Andreev, 2011). However, these canals have only been shown to exist in the incisiform teeth of the latter, whereas they are present in both the incisiform and the grinding teeth of *Sargodon tomicus.* Since pycnodonts and *Sargodon tomicus* are not closely related, and because this acrodin layout has not been detected in any other actinopterygian to date, it is more prudent to consider that this type of tissue appeared convergently within these two groups. Pycnodonts, unlike most other actinopterygians, have a very limited tooth regeneration capacity, and the same may also have been the case for *Sargodon tomicus.* Thus, it has been suggested that the odontoblasts present in the canals may have been able to maintain the acrodine cap throughout the lifetime of the animal (Andreev, 2011). However, canals penetrating the lower part of the enameloid are also known to exist in some hybodonts with grinding dentition, such as *Asteracanthus* and *Heteroptychodus* (see Chapter 3). However, these canals do not penetrate into the enameloid as deeply as in *Sargodon tomicus,* possibly excepting the case of the selachimorph shark *Ptychodus* (Figure 6.6). Nevertheless, their presence in sharks, the teeth of which must be regenerated due to the cartilaginous nature of their jaws, seems to negate a possible correlation between the possession of such structures and a particularly long tooth renewal process.

In hybodonts, these canals have been considered more as systems helping to dissipate compressive forces to which the teeth were subjected, and are always associated with a TBE, as in *Sargodon* and the pycnodonts (see Chapter 3). This type of function may also be applicable to actinopterygian teeth, but here again, the lack of experimental data does not allow rigorous testing of this theory. Moreover, it does not explain the presence of these canals within incisiform teeth. Nevertheless, the appearance of acrodin penetrated by canals originating from the dentin, with its convergent appearance in two groups of actinopterygians with grinding teeth, is probably the result of an adaptation to similar mechanical stresses rather than a reflection of their common evolutionary history.

During his work on the microstructure of the acrodin of *Sargodon tomicus* and pycnodonts, Plamen Andreev also showed the presence of a mineralized basal lamina at the junction between the acrodin and the dentin in *Sargodon*. This basal lamina marks the separation between the epithelium and the mesenchyme. In modern actinopterygians, where the acrodin mineralizes in a centripetal manner, it is located above the acrodin, suggesting that this acrodin forms in a centrifugal manner in *Sargodon tomicus*, that is, in the same manner as ganoin. This is not the case with the pycnodonts, and the hypothesis of centrifugal acrodin formation in *Sargodon tomicus* remains, for now, an anomaly that is difficult to explain. It is possible that these observations correspond to a pathological condition in the teeth studied.

In conclusion, it is undeniable that actinopterygian acrodin structure shows striking similarities to the enameloid layout found in some elasmobranchs. Of particular note are the presence of an outer PBE and an inner TBE and a similar change in the PBE bundle orientation in tooth cutting edges. These similarities have led many authors to support the hypothesis of a common evolutionary origin of these tissues. The differences shown by other studies, particularly with regard to the composition of the matrix of these tissues, have then been interpreted as resulting from a different evolutionary history for each of these groups.

However, these differences have pushed other authors, like the Japanese researcher Ichiro Sasagawa in particular, to suggest that these similarities are rather the product of an evolutionary convergence (Sasagawa, 1993). Thanks to recent research and discoveries, this interpretation is now supported by the fossil record and the developmental data. We have already seen in the preceding chapters that some of the earliest known taxa, distributed at the base of the chondrichthyans and the actinopterygians, show teeth that lack these hypermineralized tissues. Although this fact remains difficult to understand from a functional perspective, it creates doubt with regard to the homology of these two tissues. Research in developmental biology should in the near future allow us to obtain a definitive answer to this question. It has indeed been shown in teleosts as well as in an amphibian species that has enameloid on its teeth during the larval stage that certain genes involved in the formation and mineralization of the matrix of these hypermineralized tissues are co-expressed by the odontoblasts and the ameloblasts. Conversely, preliminary data obtained for two species of chondrichthyans, the small-spotted catshark (*Scyliorhinus canicula*) and the thornback ray (*Raja clavata*), indicate that although these same genes are expressed by the odontoblasts, they are not expressed by the ameloblasts, which again raises the question of the involvement of the latter in enameloid formation.

While we await the publication of these results, the question therefore remains unresolved; however, these data support the hypothesis that enameloid, in the broad sense of the term, designates a group of tissues convergent in osteichthyans and chondrichthyans. The similarities described are only morphologically superficial and due perhaps to comparable biomechanical stresses. They are not the product of a common evolutionary history. Answers may also be found by studying the fossil groups basal to the osteichthyans and the chondrichthyans: placoderms, or armored fish, known to exist from the Silurian to the Devonian (443–359 million years ago), and the acanthodians, or spiny "sharks", found from the Silurian to the Permian (443–252 million years ago). What can we learn about the presence of enameloid and its structure in these animals?

Placoderms have long been considered the most basal jawed vertebrates, forming a clade representing the sister-group of all other gnathostomes. Coming next, from most basal to most derived, are the chondrichthyans, the acanthodians and the osteichthyans (Figure 6.7A). However, recent research has cast serious doubts on this phylogenetic hypothesis. First, placoderms do not form a clade but a paraphyletic group including at least three clades (antiarchs, ptyctodonts and arthrodires), each with its own evolutionary history (Donoghue and Keating, 2014, but see King *et al.*, 2016 for a different opinion). Next, there is at least one placoderm, called *Entelognathus*, which now seems closer to the osteichthyans (bony fish) than to the chondrichthyans (Zhu *et al.*, 2013). The phylogenetic relationships of placoderms with one another as well as with the other major groups of gnathostomes appear much more complex today than they have been in the past (King *et al.*, 2016) (Figure 6.7B).

Unfortunately, placoderms are not suited for the study of hypermineralized tissues, as their jaws generally do not possess teeth, but rather gnathal plates composed of a specific form of dentine called semidentin (Benton, 2015). The presence of true teeth has been noted in some basal forms however, both in the pharyngeal cavity and on some gnathal plates. Some of these teeth are covered with a non-bundled enameloid, which would tend to prove that SCE appears before BCE and that the enameloid loss is secondary in placoderms (Rücklin and Donoghue, 2015). This idea remains the subject of debate within the scientific community (Burrow *et al.*, 2016; Rücklin and Donoghue, 2016), but the absence of enameloid in derived placoderms, whether original or secondary, would explain the fact that basal osteichthyans lacked enameloid. This would mean that it was actually a characteristic inherited from the common ancestor they shared with *Entelognathus*. The result of this would be that ganoin and acrodin are neoformations and are not homologous either in part or in total to the chondrichthyan enameloid. However, we still do not know whether the very first gnathostomes possessed teeth covered with hypermineralized tissue.

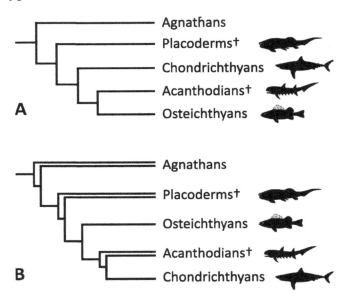

Figure 6.7. *A: Old hypothesis concerning the phylogenetic relationships of basal vertebrates. B: Recent hypothesis. The double lines indicate paraphyletic groups*

Concerning acanthodians, they are no longer considered to represent an intermediate group between the chondrichthyans and the osteichthyans, but rather to be a chondrichthyan stem group (Zhu *et al.*, 2013). This means that they have no phylogenetic relationship to the osteichthyans (Figure 6.7). Like the placoderms, they form a paraphyletic group, including at least two clades. The teeth tend to disappear in many lineages, and data concerning their hypermineralized tissues are therefore rare. In spite of this, they are generally considered to lack enamel- or enameloid-type hypermineralized tissues (Donoghue and Sansom, 2002; Andreev *et al.*, 2016). Again, this suggests that chondrichthyan enameloid, like osteichthyan ganoin and acrodin, is a neoformation appearing within a lineage originally lacking hypermineralized dental tissue.

Although the placoderms and the acanthodians, due to their stratigraphic distribution restricted to the Palaeozoic and their unfortunate tendency to lack teeth, remain very difficult to study, all of the available data have converged to indicate that acrodin and ganoin, on one hand, and chondrichthyan enameloid, on the other hand, appeared convergently. Another characteristic suggesting that these tissues are not homologous is their mineralogic composition: fluorapatite for chondrichthyans, and hydroxyapatite with a variable amount of fluorine for osteichthyans. In view of the significant similarities in the microstructures of acrodin and BCE, it would seem, therefore, that the evolution of these tissues was essentially governed by their mechanical properties. A TBE-type structure is systematically selected in grinding teeth, whereas a PBE-type structure is favored in the pointed teeth of predators that consumed soft-bodied prey. It is important to note, however, that even if the hypermineralized tissues of chondrichthyans and osteichthyans are the result of convergence, they develop following the same pattern, with an SCE-type tissue appearing before a BCE-type one. Placoderms might be able to confirm this hypothesis if the presence of non-bundled enameloid is confirmed in certain basal forms, which may indicate that the formation of hypermineralized tissue is automatically initiated by epithelial cells and that the involvement of mesenchymal cells can only be secondary. This is a fascinating research topic that will require further study.

Conclusion

Studying the evolutionary history of selachians is a complicated task. The simple fact that they possess a cartilaginous skeleton makes their fossilization difficult except under exceptional conditions, as cartilage deteriorates much more rapidly than bone. As noted in the introduction, complete fossils are, therefore, extremely rare. On the other hand, during their lifetime, most selachians produce large numbers of teeth that can reach into the thousands because of their continuous dental regeneration, and the highly mineralized nature of these teeth facilitates their fossilization. The vast majority of fossil selachian species is thus known only from isolated teeth and reconstructing the history of such a diversified group on the basis of such fragmentary material is not an easy task. This makes any supplementary information about their dental morphology very welcome, and the microstructure of hypermineralized tissues remains an invaluable tool.

As we have seen in this book, microstructural study was initially aimed mainly at identifying the most ancient selachimorphs in the fossil record, thanks to the complexity of their enameloid, in order to better understand the timing of their appearance. However, the consequences of this relatively limited use proved to be relatively harmful:

– These studies focused on selachimorphs, and the other selachian groups were less studied and sometimes even excluded altogether.

– They favored the simplistic idea that the enameloid microstructure could be divided into two groups: complex (with three "layers", SLE, PBE and TBE) for selachimorphs and simple (with one layer, SCE) for all the other selachians.

– Since the "utilitarian" application of the method was aimed simply at showing the presence of PBE in order to demonstrate the studied tooth belonging (or not) to the selachimorphs, it led to the favoring, for the sake of convenience, of surface studies, which were simpler to conduct. However, they did not yield as detailed an understanding of microstructure as the studies conducted using cross-sections.

It was not until the last decade that the real diversity of enameloid microstructures finally began to be recognized. Enameloid is now divided into two main units: SCE and BCE. The latter shows three components: TBE, RBE and PBE (see Chapter 1). The tissue can then be broken down into a large number of combinations of units and components (see previous chapters). However, it is clear that despite the efforts made during the last few decades, the sampling of living and fossil species taken from among the chondrichthyans for histological use remains very sparse compared to the diversity of this group. There is still a great deal to be done in terms of examining the dental microstructure of representatives of the major clades of chondrichthyans (Paleozoic groups, holocephali, hybodonts, selachimorphs and batomorphs). However, a simple increase in sampling will not be enough to cast light on the gray areas that are still present in the distribution of histological characteristics in the chondrichthyan family tree. Our understanding of the polarity of enameloid characters is dependent on our understanding of the phylogenetic relationships among the different groups. Yet, these relationships remain poorly understood, particularly with regard to the various clades of chondrichthyans that lived during the Paleozoic Era, as well as to certain neoselachian groups.

But we have made progress, and our greater understanding of the enameloid diversity has brought its own questions, the most important of which is undoubtedly whether or not these different microstructures reflect a phylogenetic signal or simply a functional adaptation. In other words, does it provide reliable information that will help us to study the phylogenetic relationships of selachians, or does its adaptation to the mechanical function of the tooth completely blur this phylogenetic signal? As we have seen in the previous chapters, the structural adaptation to the mechanical stresses undergone by the tooth is undeniable, but despite this, the phylogenetic signal seems preserved in most cases. Selachimorphs and some ctenacanths share a PBE which helps their teeth to resist tensile forces more effectively; however, the organizational details of this PBE show significant variations that enable us to differentiate between the teeth of these two groups (see Chapters 2 and 5). The stability of the enameloid microstructure of myliobatiform teeth, regardless of their diet, is another example of the preservation of phylogenetic information conveyed by enameloid microstructure (see Chapter 4).

Still, we must remain cautious. The troubling similarities between the general organization of the BCE in the tooth enameloid of selachimorphs and the acrodin in the teeth of actinopterygians may suggest that these tissues are homologous, i.e. they appeared in the common ancestor shared by these two lineages. It is important to remember here that acrodine lacks the equivalent of SCE, and thus cannot be homologous to the entirety of selachian enameloid, unless it has lost its SCE unit secondarily. Yet, on closer examination, we can see major differences between BCE and acrodine. For example, the composition of their mineral phase is different, i.e. hydroxyapatite with a variable fluorine content for the latter and fluorapatite for the former (Berkovitz and Shellis, 2017). The composition of their protein matrix is also different (see Chapter 6). Thus, a more careful way of considering it is that their microstructural similarities are the result of convergent evolution, with the tissues independently exhibiting the same organization as an adaptation to the same types of mechanical stresses. An estimation of the real impact of these phenomena on the evolution of these tissues remains difficult to make, however, due to the lack of reliable mechanical data. It is true that studies of the mechanical resistance of selachimorph teeth have been carried out; however, they have only considered enameloid as a whole (Enax *et al.*, 2014), rather than looking at its component parts. It is generally thought, as we have reiterated many times in the pages of this book, that PBE strengthens the teeth against tensile forces, whereas TBE enables a more effective dissipation of compressive forces. However, this has never been quantified, and the evolutionary benefit, in terms of natural selection, of each of these components in relation to a given diet cannot be understood with precision. It would be of crucial importance to test certain functional hypotheses within a strict experimental framework, perhaps using computer models, in order to confirm the role of certain types of microstructure in the context of resistance to torsion or crushing. The next stage in the study of the mechanisms of enameloid evolution will thus be developing models that can be used to measure and quantify the mechanical characteristics of each unit (SCE and BCE) and of each component (TBE, RBE and PBE), as well as the different combinations of these. When a tool of this type is available, we will truly be able to assess their importance in terms of adaptation to a given diet, and thus to better assess their probability of being retained via natural selection. This would help us to understand to what extent mechanical properties influence the evolution of the tissue. Such an approach would probably require close cooperation with material physicists and specialists in biomechanics, which would bring new innovative perspectives to the study of extinct organisms.

However, this purely functional approach will not be sufficient. We must also identify the genes responsible for the production of these structures in order to follow their distribution within the various lineages. Of course, this will only be possible for lineages still possessing living representatives, but this stage is vital to our understanding of how a microstructural pattern arises and subsequently endures, regardless of selective stresses related to its mechanical properties. For example, does the maintenance of a relatively non-structured enameloid in the Myliobatiformes correspond to the loss of certain genes (or changes in their activity and regulation) within that order? Moreover, it will help us to understand whether the same genes, and thus the same proteins, are responsible for the production of BCE and acrodine. If this is not the case, the fact that they are not homologous tissues will be proven.

This leads us to wonder about the nature of the first hypermineralized tissues in gnathostomes. As we saw in Chapter 2, the most basal selachian in which the enameloid microstructure has been studied, *Leonodus*, seems to possess only an SCE, which appears to be confirmed by the study of dermal denticles (see Chapter 2). This may indicate that the primitive condition for selachians is a single-unit enameloid. In this case, the appearance of the second unit, the BCE, will be taking place later on, though still rapidly, because a "proto-BCE" is present in taxa that are only slightly more derived than *Leonodus*, such as *Portalodus*. However, as we have emphasized, the study of the microstructure of these ancient teeth, which are also extremely small, remains technically complicated, and it is sometimes difficult to separate real microstructure from preservation artifacts.

We are also confronted with a mystery: the presence of basal selachians whose teeth lack enameloid, including all the Phoebodontiformes and the Xenacanthimorphs (see Chapter 2). Once again, we lack a mechanical model to assess the degree to which the absence of the hypermineralized covering impacted the mechanical resistance of the tooth. If this impact was low, the hypothesis of a secondary loss remains credible. But if the impact was high, it becomes more difficult to explain. Can we really be sure, given the small number of teeth studied to date, that the plesiomorphic state for selachian teeth is the presence of a SCE, and not the absence of enameloid?

We can also consider that our sampling reflects reality, and that SCE is in fact the plesiomorphic state for selachians with a subsequent appearance of BCE, and sometimes a secondary disappearance of enameloid. How does this scenario compare with what has been observed in actinopterygians? If we consider, based solely on their similarities in terms of organization, that SCE is equivalent to ganoine and BCE to acrodine, we can see a certain similarity as in actinopterygians, ganoine appears before acrodine (see Chapter 6). However, one of the oldest bony fish known, *Andreolepis*, possesses teeth lacking any hypermineralized tissue. The appearance of this type of tissue in the osteichthyans would thus be secondary, which is a scenario that differs greatly from the hypothesis which we have accepted as the most probable for selachians, in which the oldest teeth are already covered with a "simple" SCE-type enameloid. Does this mean that tooth tissue, or even teeth, evolved in a radically different manner in osteichthyans and selachians?

Finally, the study of the tooth enameloid microstructure in basal gnathostomes reveals a complex history of the teeth within this lineage. It now appears virtually certain that different enameloids appeared multiple times independently of one another, and are not homologous to each other. In this case, what about the teeth themselves? Their internal structure is remarkably stable, built around a pulp cavity surrounded by dentin, but is this enough to demonstrate that the tooth structure appeared only once in the history of the gnathostomes? Could the teeth of chondrichthyans and osteichthyans also be the result of convergence? In this case, what is the link between the appearance of jaws and the appearance of teeth? It is counter-intuitive to dissociate the two, but if the teeth are not in fact homologous from one lineage to the other, their appearance cannot really be linked to that of jaws any longer. In addition to their practical interest for the identification of the phylogenetic affinities of fossil teeth found isolated, the study of the microstructure of hypermineralized tissues may lead us to reconsider our understanding of a major event in the history of vertebrates: the appearance of teeth and jaws.

Glossary

Acrodine: a type of hypermineralized tissue formed of bundles of crystals, covering the apical part of the teeth of actinopterygians. It is a type of enameloid.

Actinopterygians: ray-finned bony fishes.

Agnathans: vertebrates characterized by a jawless mouth.

Ameloblasts: prismatic cells derived from the dental epithelium, responsible for the formation of tooth enamel.

Batomorphs: clade of chondrichthyans, including both fossil and modern skates and rays.

BCE (Bundled Crystallite Enameloid): enameloid unit made up of three distinct components (PBE, TBE and RBE) containing bundled crystals.

Cenozoic: a geologic era beginning after the Cretaceous/Paleogene extinction event (66 mya) and continuing to the present day.

Chimaera (Chimaeriformes): group of cartilaginous fish and sister group of the elasmobranchs.

Chondrichthyans: clade including both fossil and modern cartilaginous fishes (elasmobranchs and chimaera).

Clade: monophyletic group including all the descendants of a common ancestor.

Collagen: family of proteins constituting the largest part of the extracellular matrix of animal tissues.

Dentine: mineralized hydroxyapatite tissue forming the root and crown of vertebrate teeth and some scales.

Durophage (n) / durophagous (adj): organisms whose diet is composed of hard-shelled preys that must be crushed, such as shelled invertebrates or crustaceans.

Ecological niche: place occupied by a species within an ecosystem.

Elasmobranchs: sub-class of Chondrichthyes including sharks, skates and rays (neoselachians) as well as their fossil sister group (hybodonts).

Enameloid: hypermineralized tissue covering the surface of the tooth crowns of chondrichthyans and actinopterygians as well as the scales and spines of chondrichthyans, which is formed of hydroxyapatite or fluorapatite crystals.

EMP (Extracellular Matrix Proteins): group of proteins characteristic of enamel, which are secreted by ameloblasts during the tooth development.

Epithelium: more or less complex tissue formed of cells closely juxtaposed with one another, carrying out the functions of protective coating and substance synthesis (glandular function).

Ganoine: a type of hypermineralized tissue formed of stratified microcrystals, covering some scales and the lower part of the teeth of actinopterygians. This tissue is considered homologous with enamel.

Gnathostomes: vertebrates characterized by a mouth with jaws.

Heterodonty: variation of dental morphology within a jaw (monognathic), between two jaws (dignathic), among individuals of different sexes (gynandric) or across time (ontogenic).

Histology: field of biology studying the composition and morphology of organic tissues.

Homology: similarity of characteristics (anatomic and molecular) present in two taxa due to the inheritance of this characteristic from a common ancestor.

Hybodonts: clade of chondrichthyans that disappeared at the end of the Cretaceous period and the sister group of neoselachians.

Labial: on the outside of the jaw.

Lingual: on the inside of the jaw.

LRB (Loose Radial Bundles): radial enameloid microcrystals originating in SCE and penetrating PBE bundles in a perpendicular fashion.

Mesenchyme: embryonic support tissue usually originating in the mesoderm (more rarely in the ectoderm).

Mesozoic: a geologic era extending from the Triassic (252 mya) to the Cretaceous/ Paleogene boundary (66 mya).

Microphage: organism whose diet is composed of small particles, generally ingested via filtration.

Microstructure: micrometric crystalline structure of mineralized tissue.

Monophyletic: designates a clade in which all the representatives share a common ancestor.

Morphological convergence: evolutionary mechanism leading to morphological resemblance acquired independently by two non-related taxa in response to similar environmental stresses.

Neoselachians: clade of chondrichthyans, including batomorphs and selachimorphs.

Odontoblasts: specialized cells deriving from the dental mesenchyme; they are responsible for the formation of dentine.

Odontogenesis: the development of teeth.

Ontogenesis: the development of different stages of an organism from its conception to the adult state.

Orthodentine: mineralized tissues around the pulp cavity, forming the crown of teeth and the dermal denticles of certain chondrichthyans.

Osteichthyans: clade including all bony fishes as well as the tetrapods.

Osteodentine: trabecular bone forming the root and, in some taxa, filling in the pulp cavity of the tooth crown.

Paleozoic: a geologic era extending from the Cambrian (541 mya) to the Permian/Triassic boundary (–252.2 mya).

Paraphyletic: describes a group in which the taxa represent only a portion of the descendants of a common ancestor.

PBE (Parallel Bundled Enameloid): enameloid component in which the crystals are arranged in bundles that are more or less compact and parallel to the crown surface.

Phylogenetics: the discipline that studies the evolutionary history and relationships between taxa.

Planktonophage: organism whose diet is composed of plankton.

Pleromine: a type of hypermineralized tissue present in the dental plates of chimaera, the nature and origin of which remains poorly understood.

Plesiomorphic: describes an ancestral characteristic that has remained unchanged throughout evolution.

RBE (Radial Bundled Enameloid): enameloid component composed of radial bundles of microcrystals oriented perpendicular to the tooth surface.

SCE (Single Crystallite Enameloid): enameloid unit composed of isolated and randomly oriented microcrystals.

Selachimorphs: clade of chondrichthyans, including fossil and modern neoselachian sharks.

SLE (Shiny Layer Enameloid): enameloid component composed of individualized and randomly oriented fluorapatite crystals. This name is no longer used for this component and must be replaced by SCE.

Stratigraphy: the discipline that studies the spatial and temporal succession of geological layers.

Systematics: the discipline that aims at classifying taxa.

Taxon: a systematic classification of entity (e.g. species, genus, family and order) that includes organisms sharing common characteristics.

TBE (Tangled Bundled Enameloid): enameloid component in which the bundles of crystals are entangled.

Teleosteans: most diverse group of actinopterygians characterized by mobile maxillary and premaxillary bones.

Tetrapods: clade including all the four-limbed vertebrates.

TRB (Thick Radial Bundles): radial enameloid microcrystals grouped into bundles penetrating PBE bundles perpendicularly.

Vertebrates: group of animals possessing an internal, more or less mineralized, skeleton.

Bibliography

ADNET S., CAPPETTA H., GUINOT G. *et al.* (2012), "Evolutionary history of the devilrays (Chondrichthyes: Myliobatiformes) from fossil and morphological inference", *Zoological Journal of the Linnean Society*, vol. 166, pp. 132–159.

AGASSIZ L. (1833–1844), *Recherches sur les poissons fossiles*, Petitpierre, Neuchatel.

ANDREEV P.S. (2010), "Enameloid microstructure of the serrated cutting edges in certain fossil carcharhiniform and lamniform sharks", *Microscopy Research and Technique*, vol. 73, pp. 704–713.

ANDREEV P.S. (2011), "Convergence in dental histology between the late Triassic semionotiform *Sargodon tomicus* (Neopterygii) and a Late Cretaceous (Turonian) pycnodontid (Neopterygii: Pycnodontiformes) species", *Microscopy Research and Technique*, vol. 74, no. 5, pp. 464–479.

ANDREEV P.S., COATES M.I., KARATAJŪTĖ-TALIMAA V. *et al.* (2016), "The systematics of the Mongolepidida (Chondrichthyes) and the Ordovician origins of the clade", *PeerJ*, vol. 4, no. e1850, pp. 1–38, 2016. doi: 10.7717/peerJ.1850.

ASSARAF-WEILL N., GASSE B., SILVENT J. *et al.* (2014), "Ameloblasts express type I collagen during amelogenesis", *Journal of Dental Research*, vol. 93, pp. 502–507.

ASCHLIMAN N.C., NISHIDA M., MIYA M. *et al.* (2012), "Body plan convergence in the evolution of skates and rays (Chondrichthyes: Batoidea)", *Molecular Phylogenetics and Evolution*, vol. 63, pp. 28–42.

BENDIX-ALMGREEN S.E. (1994), "*Adamantina benedictae* n.g. et sp.- a new elasmobranch from the marine Upper Permian of East Greenland", *Ichthyolith Issues*, vol. 14, pp. 21–22.

BENTON M. (2015), *Vertebrate Palaeontology*, 4th ed., John Wiley & Sons, Chichester.

BERKOVITZ B., SHELLIS P. (2017), *The Teeth of Non-mammalian Vertebrates*, Academic Press, London.

BLAIS S.A., MACKENZIE L.A., WILSON M.V.H. (2011), "Tooth-like scales in Early Devonian eugnathostomes and the 'outside-in' hypothesis for the origins of teeth in vertebrates", *Journal of Vertebrate Paleontology*, vol. 31, no. 6, pp. 1189–1199.

BLAZEJOWSKI B. (2004), "Shark teeth from the Lower Triassic of Spitsbergen and their histology", *Polish Polar Research*, vol. 25, pp. 153–167.

BOTELLA H., DONOGHUE P.C.J., MARTÍNEZ-PÉREZ C. (2009), "Enameloid microstructure in the oldest known chondrichthyan teeth", *Acta Zoologica*, vol. 90, pp. 103–108.

BURROW C.J., HOVESTADT D.C., HOVESTADT-EULER M. *et al.* (2008), "New information on the Devonian shark *Mcmurdodus*, based on material from western Queensland, Australia", *Acta Geologica Polonica*, vol. 58, no. 2, pp. 155–163.

BURROW C.J., HU Y., YOUNG G. (2016), "Placoderm and the evolutionary origin of teeth: a comment on Rücklin and Donoghue (2015)", *Biology Letters*, vol. 12, no. 9. doi: 10.1098/rsbl.2016.0159.

CANAVARI M. (1916), "Descrizione di un notevole esemplare di *Ptychodus* Agassiz trovato nel calcare bianco della creta superiore di Gallio nei sette communi (Veneto)", *Palaeontographia italica*, vol. 22, pp. 35–102.

CAPPETTA H. (2012), *Chondrichthyes. Mesozoic and Cenozoic Elasmobranchii: Teeth*, Verlag Dr. Friedrich Pfeil, München.

CHEN C., WANG Z., SAITO M. *et al.* (2014), "Fluorine in shark teeth: its direct atomic-resolution imaging and strengthening function", *Angewandte Chemie-International Edition*, vol. 53, pp. 1543–1547.

CHEN D., BLOM H., SANCHEZ S. *et al.* (2016), "The stem osteichthyan *Andreolepis* and the origin of tooth replacement", *Nature*, vol. 539, pp. 237–241.

COMPAGNO L.V.J. (1990), "Alternative life-history styles of cartilaginous fishes in time and space", *Environmental Biology of Fishes*, vol. 28, pp. 33–75.

CUNY G. (1998), "Primitive neoselachian sharks: a survey", *Oryctos*, vol. 1, pp. 3–21.

CUNY G., BENETEAU A. (2013), *Requins, de la préhistoire à nos jours*, Belin, Paris.

CUNY G., BENTON M.J. (1999), "Early radiation of the neoselachian sharks in Western Europe", *Geobios*, vol. 32, pp. 193–204.

CUNY G., COBBETT A.M., MEUNIER F.J. *et al.* (2010), "Vertebrate microremains from the Early Cretaceous of southern Tunisia", *Geobios*, vol. 43, pp. 615–628.

CUNY G., HUNT A., MAZIN J.-M. *et al.* (2000), "Teeth of enigmatic neoselachian sharks and an ornithischian dinosaur from the uppermost Triassic of Lons-le-Saunier (Jura, France)", *Paläontologische Zeitschrift*, vol. 74, pp. 171–185.

CUNY G., MARTIN M., RAUSCHER R. *et al.* (1998), "A new neoselachian shark from the Upper Triassic of Grozon (Jura, France)", *Geological Magazine*, vol. 135, pp. 657–668.

CUNY G., RISNES S. (2005), "The enameloid microstructure of the teeth of Synechodontiform sharks (Chondrichthyes: Neoselachii)", *PalArch*, vol. 3, no. 2, pp. 8–19.

CUNY G., SRISUK P., KHAMHA V. *et al.* (2009), *A New Elasmobranch Fauna from the Middle Jurassic of Southern Thailand*, Geological Society, London, Special Publications 315, pp. 97–113.

DE CARVALLO M.R. (1996), "Higher-level elasmobranch phylogeny, basal Squaleans, and paraphyly", in STIASSNY M.L.J., PARENTI L.R., JOHNSON G.D. (eds), *Interrelation of Fishes*, Academic Press, New York, pp. 35–62.

DEAN B. (1909), "Studies of fossil fishes (sharks, chimaeroids and arthrodires)", *Memoirs of the American Museum of Natural History*, Part V, vol. 9, pp. 211–287.

DEAN M.N., BIZZARO J.J., SUMMERS A.P. (2007), "The evolution of cranial design, diet and feeding mechanisms in batoid fishes", *Integrative and Comparative Biology*, vol. 47, pp. 70–81.

DEBIAIS-THIBAUD M., OULION S., BOURRAT F. *et al.* (2011), "The homology of odontodes in gnathostomes: insights from Dlx gene expression in the dogfish, *Scyliorhinus canicula*", *BMC Evolutionary Biology*, vol. 11, p. 307.

DELSATE D., CANDONI L. (2001), "Description de nouveaux morphotypes dentaires de Batomorphii toarciens (Jurassique inférieur) du Bassin de Paris Archaeobatidae nov", *Fam. Bulletin de la Societé des Naturalistes Luxembourgeois*, vol. 102, pp. 131–143.

DINGERKUS G., SÉRET B., GUILERT E. (1991), "Multiple prismatic calcium phosphate layers in the jaw of present-day sharks (Chondrichthyes; Selachii)", *Experientia*, vol. 47, pp. 38–40.

DONOGHUE P.C.J., SANSOM I.J. (2002), "Origin and early evolution of vertebrate skeletonization", *Microscopy Research and Technique*, vol. 59, pp. 352–372.

DONOGHUE P.C.J., SANSOM I.J., DOWNS J.P. (2006), "Early evolution of vertebrate skeletal tissues and cellular interactions, and the canalization of skeletal development", *Journal of Experimental Zoology*, vol. 306B, pp. 278–294.

DONOGHUE P.C.J., KEATING J.N. (2014), "Early vertebrate evolution", *Palaeontology,* vol. 57, no. 5, pp. 879–893.

DOUADY C.J., DOSAY M., SHIVJI M.S. *et al.* (2003), "Molecular phylogenetic evidence refuting the hypothesis of Batoidea (rays and skates) as derived sharks", *Molecular Phylogenetics and Evolution*, vol. 26, pp. 215–221.

DUFFIN C.J. (1980), "A new euselachian shark from the Upper Triassic of Germany", *Neues Jahrbuch für Geologie und Paläontologie, Monatshefte*, vol. 1980, pp. 1–16.

DUFFIN C.J. (1981), "Comments on the selachian genus *Doratodus* SCHMID, 1861 (Upper Triassic, Germany)", *Neues Jahrbuch für Geologie und Paläontologie, Monatshefte*, vol. 1981, pp. 289–302.

DUFFIN C.J. (1998), "New shark remains from the British Rhaetian (latest Triassic). 2. Hybodonts and palaeospinacids", *Neues Jahrbuch für Geologie und Paläontologie, Monatshefte*, vol. 1998, pp. 240–256.

DUFFIN C.J. (2001), "The hybodont shark, *Priohybodus* d'Erasmo, 1960 (Early Cretaceous, northern Africa)", *Zoological Journal of the Linnean Society,* vol. 133, pp. 303–308.

DUFFIN C.J., CUNY G. (2008), "*Carcharopsis prototypus* and the adaptations of single crystallite enameloid in cutting dentitions", *Acta Geologica Polonica,* vol. 58, no. 2, pp. 181–184.

DUFFIN C.J., WARD D.J. (1983), "Neoselachian shark's teeth from the Lower Carboniferous of Britain and the Lower Permian of the U.S.A", *Palaeontology*, vol. 26, no. 1, pp. 93–110.

DUNN K.A., MCEACHRAN J.D., HONEYCUTT R.L. (2003), "Molecular phylogenetics of myliobatiform fishes (Chondrichthyes: Myliobatiformes), with comments on the effects of missing data on parsimony and likehood", *Molecular Phylogenetics and Evolution*, vol. 27, pp. 259–270.

EBERT D.A., FOWLER S., COMPAGNO L. (2013), *Sharks of the World*, Wild Nature Press, Plymouth.

ENAULT S., CAPPETTA H., ADNET S. (2013), "Simplification of the enameloid microstructure of large stingrays (Chondrichthyes: Myliobatiformes): a functional approach", *Zoological Journal of the Linnean Society,* vol. 169, pp. 144–155.

ENAULT S., GUINOT G., MARTHA B. *et al.* (2015), "Chondrichthyan tooth enameloid: past, present and future", *Zoological Journal of the Linnean Society,* vol. 174, pp. 549–570.

ENAX J., JANUS A.M., RAABE D. *et al.* (2014), "Ultrastructural organization and micromechanical properties of shark tooth enameloid", *Acta Biomaterialia,* vol. 10, no. 9, pp. 3959–3968.

FRANCILLON-VIEILLOT H., DE BUFFRÉNIL V., CASTANET J. *et al.* (1990), "Microstructure and mineralization of vertebrate skeletal tissues", in CARTER J.G. (ed.), *Skeletal Biomineralization: Patterns, Processes and Evolutionary Trends*, Van Nostrand Reinhold, New York, pp. 471–530.

GARDINER B.G., SCHAEFFER B., MASSERIE J.A. (2005), "A review of the lower actinopterygian phylogeny", *Zoological Journal of the Linnean Society,* vol. 144, no. 4, pp. 511–525.

GILLIS J.A., DONOGHUE P.C.J. (2007), "The homology and phylogeny of chondrichthyan tooth enameloid", *Journal of Morphology,* vol. 268, pp. 33–49.

GINTER M., HAMPE O., DUFFIN C. (2010), *Chondrichthyes. Palaeozoic Elasmobranchii: Teeth,* Verlag Dr. Friedrich Pfeil, München.

GLIKMAN L.S. (1964), *Sharks of Paleogene and their Stratigraphic Significance*, Nauka Press, Moscow.

GOTO M. (1978), "Histogenetic studies on the teeth of leopard shark (Triakis scyllia)", *Journal of the Stomatological Society*, Japan, vol. 45, pp. 527–584.

GRADY J.E. (1970), "Tooth development in sharks", *Archives of Oral Biology,* vol. 15, pp. 613–619.

GUINOT G., ADNET S., CAVIN L. *et al.* (2013), "Cretaceous stem chondrichthyans survived the end-Permian mass extinction", *Nature Communications,* vol. 4, no. 2669, pp. 1–8.

GUINOT G., CAPPETTA H. (2011), "Enameloid microstructure of some Cretaceous Hexanchiformes and Synechodontiformes (Chondrichthyes, Neoselachii): new structures and systematic implications", *Microscopy Research and Technique*, vol. 74, pp. 196–205.

HALL B.K. (2005), *Bones and Cartilage: Developmental and Evolutionary Skeletal Biology*, Elsevier Academic Press, Sans Diego, California & London.

HAMPE O. (1991), "Histological investigations on fossil teeth of the shark Order Xenacanthida (Chonfrichthyes: Elasmobranchii) as revealed by fluorescence microscopy", *Scientific and Technical Information*, vol. 10, no. 1, pp. 17–27.

HAMPE O., LONG J.A. (1999), "The histology of middle Devonian chondrichthyan teeth from southern Victoria Land, Antarctica", *Records of the Australian Museum Supplement*, vol. 57, pp. 23–36.

HENRIKSEN N. (2008), *Geological History of Greenland*, GEUS, Copenhagen.

HEROLD R.C., GRAVER H.T., CHRISTNER P. (1980), "Immuno-histochemical localization of amelogenins in enameloid of lower vertebrate teeth", *Science,* vol. 207, pp. 1357–1358.

HODNETT J.-P., ELLIOTT D.K., OLSON T.J. *et al.* (2012), "Ctenacanthiform sharks from the Permian Kaibab Formation, northern Arizona", *Historical Biology*, vol. 24, no. 4, pp. 381–395.

HODNETT J.-P., ELLIOTT D.K., OLSON T.J. (2013), "A new basal hybodont (Chondrichthyes, Hybodontiformes) from the Middle Permian (Roadian) Kaibab Formation of northern Arizona", *New Mexico Museum of Natural History and Science Bulletin,* vol. 60, pp. 103–108.

HOFFMAN B.L., HAGEMAN S.A., CLAYCOMB G.D. (2016), "Scanning electron microscope examination of the dental enameloid of the Cretaceous durophagous shark *Ptychodus* supports neoselachian classification", *Journal of Paleontology,* vol. 90, no. 4, pp. 741–762.

HUYSSEUNE A., TAKLE H., SOENENS M. *et al.* (2008), "Unique and shared gene expression patterns in Atlantic salmon (Salmo salar) tooth development", *Development Genes and Evolution,* vol. 218, pp. 427–437.

HUYSSEUNE A., SIRE J.-Y. (1998), "Evolution of pattern and processes in teeth and tooth-related tissues in non-mammalian vertebrates", *European Journal of Oral Sciences,* vol. 106, pp. 437–481.

ISHIYAMA M., SASAGAWA I., AKAI J. (1984), "The inorganic content of pleromin in tooth plates of the living holocephalan, Chimaera phantasma, consists of a crystalline calcium phosphate known as beta-Ca3(PO4)2 (whitlockite)", *Archives of Histology and Cytology*, vol. 47, pp. 89–94.

ISHIYAMA M., YOSHIE S., TERAKI Y. *et al.* (1991), "Ultrastructure of pleromin, a highly mineralized tissue comprizing crystalline calcium phosphate known as whitlockite, in holocephalian tooth plates", in SUGA S., NAKAHARA H. (eds), *Mechanisms and Phylogeny of Mineralization in Biological Systems*, Springer-Verlag, Tokyo, pp. 453–457.

IVANOV A. (2005), "Early Permian chondrichthyans of the middle and south Urals", *Revista Brasileira de Paleontologia*, vol. 8, pp. 127–138.

JANVIER P. (1978), On the oldest known teleostome fish *Andreolepis hedei* Gross (Ludlow of Gotland) and the systematic position of the lophosteids, *Eesti NSV Teaduste Akadeemia Toimetised Geoloogia*, vol. 27, pp. 88–95.

KAWASAKI K. (2009), "The SCPP gene repertoire in bony vertebrates and graded differences in mineralized tissues", *Development Genes and Evolution*, vol. 219, pp. 147–157.

KAWASAKI K. (2013), "Odontogenetic ameloblast-associated protein (ODAM) and amelotin: major players in hypermineralization of enamel and enameloid", *Journal of Oral Biosciences,* vol. 55, pp. 85–90.

KAWASAKI K., SUZUKI T., WEISS K.M. (2005), "Phenogenetic drift in evolution: the changing genetic basis of vertebrate teeth", *Proceedings of the National Academy of Sciences of the USA*, vol. 102, pp. 18063–18068.

KEMP N. (1985), "Ameloblastic secretion and calcification of the enamel layer in shark teeth", *Journal of Morphology*, vol. 184, pp. 215–230.

KEMP N. (1999), "Integumentary system and teeth", in HAMLETT W.C. (ed.), *Shark, Skates and Rays: The Biology of Elamsoblanch Fishes*, John Hopkins Press, Baltimore, MD, pp. 43–68.

KEREBEL L.-M., LE CABELLEC M.-T. (1980), "Enameloid in the teleost fish *Lophius*", *Cell and Tissue Research*, vol. 206, pp. 211–223.

KEREBEL B., DACULSI G., RENAUDIN S. (1977), "Ultrastructure des améloblastes au cours de la formation de l'émailloïde des sélaciens", *Biologie Cellulaire,* vol. 28, pp. 125–130.

KING B., QIAO T., LEE M.S.Y. *et al.* (2016), "Bayesian morphological clock methods resurrect placodrem monophyly and reveal rapid early evolution of jawed vertebrates", *Systematic Biology*, pp. 1–18. doi: 10.1093/sysbio/syw107.

KOLMANN M.A., HUBER D.R., DEAN M.N. *et al.* (2014), "Morphological variability in a decoupled skeletal system: batoid cranial anatomy", *Journal of Morphology*, vol. 275, pp. 862–881.

KOOT M.B., CUNY G., TINTORI A. *et al.* (2013), "A new diverse fauna from the Wordian (Middle Permian) Khuff Formation in the interior Haushi-Huqf area, Sultanate of Oman", *Palaeontology*, vol. 56, no. 2, pp. 303–343.

KVAM T. (1946), "Comparative study of the ontogenetic and phylogenetic development of dental enamel", *Norske Tannlaegeforen*, vol. 56, pp. 1–198.

KVAM T. (1950), *The Development of Mesodermal Enamel on Piscine Teeth*, Trondheim Aktietrykkeriet, Trondheim, pp. 1–115.

LANDIS W.J., SILVER F.H. (2009), "Mineral deposition in the extracellular matrices of vertebrate tissues: identification of possible apatite nucleation sites on type I collagen", *Cells Tissues Organs,* vol. 189, pp. 20–24.

LAST P., WHITE W., SÉRET B. *et al.* (2016), *Rays of the World*, CSIRO Publishing, p. 801.

LEWY Z. (2002), "The function of the ammonite fluted septal margins", *Journal of Paleontology*, vol. 76, no. 1, pp. 63–69.

LÜBKE A., ENAX J., LOZA K. *et al.* (2015), "Dental lessons from past to present: ultrastructure and composition of teeth from plesiosaurs, dinosaurs, exctinct and recent sharks", *RSC Advances*, vol. 5, pp. 61612–61622.

MADER H. (1986), "Schuppen und zähne von acanthodiern und elasmobranchiern aus dem Uter-Devon Spaniens (Pises)", *Gottinger Arbeiten zur Geologie une Palaontologie*, vol. 28, pp. 1–59.

MAISEY J.G., NAYLOR G.J.P., WARD D.J. (2004), "Mesozoic elasmobranchs, neoselachian phylogeny and the rise of modern elasmobranch diversity", in ARRATIA G., TINTORI A. (eds), *Mesozoic Fishes 3 – Systematics, Paleoenvironment and Biodiversity*, Verlag Dr. Friedrich Pfeil, Munich, pp. 17–56.

MANZANARES E., PLA C., MARTÍNEZ-PÉREZ C. *et al.* (2016), "*Lonchidion derenzii*, sp. nov., a new lonchidiid shark (Chondrichthyes, Hybodontiforms) from the Upper Triassic of Spain, with remarks on lonchidiid enameloid", *Journal of Vertebrate Paleontology*. doi: 10.1080/02724634.2017.1253585.

MANZANARES E., RASSKIN-GUTMAN D., BOTELLA H. (2016), "New insights into the enameloid microstructure of batoid fishes (Chondrichthyes)", *Zoological Journal of the Linnean Society*, vol. 177, pp. 621–632.

MARSHALL A.S. (2009), "Redescription of the genus *Manta* with resurrection of *Manta alfredi* (Krefft, 1868) (Chondrichthyes: Myliobatoidei, Mobulidae)", *Zotaxa*, vol. 2301, pp. 1–28.

MCEACHRAN J.D., DUNN K.A. (1998), "Phylogenetic analysis of skates, a morphologically conservative clade of elasmobranchs (Chondrichthyes: Rajidae)", *Copeia*, vol. 2, pp. 271–290.

MOSS M.L., JONES S.J., PIEZ K.A. (1964), "Calcified ectodermal collagens of shark tooth enamel and teleost scale", *Science*, vol. 145, pp. 940–942.

MOYER J.K., RICCIO M.L., BEMIS W.E. (2015), "Development and microstructure of tooth histotypes in the blue shark, Prionace glauca (Carcharhiniformes: Carcharhinidae) and the great white shark, Carcharodon carcharias (Lamniformes: Lamnidae)", *Journal of Morphology*, vol. 276, pp. 797–817.

NOTARBARTOLO DI SCIARA G. (1987), "A revisionary study of the genus Mobula Rafinesque, 1810 (Chondrichthyes: Mobulidae) with the description of a new species", *Zoological Journal of the Linnean Society*, vol. 91, pp. 1–91.

OWEN R. (1840), *Odontography; Or, a Treatise on the Comparative Anatomy of the Teeth; Their Physiological Relations, Mode of Development, and Microscopic Structure, in the Vertebrate Animals*, Baillère H., London.

ØRVIG T. (1966), "Histologic studies of placoderms and fossil elasmobranchs. 2. On the dermal skeleton of two late Paleozoic elasmobranchs", *Arkiv för Zoologi*, vol. 2, pp. 1–39.

ØRVIG T. (1967), "Phylogeny of tooth tissues: evolution of some calcified tissues in early vertebrates", in MILES A.E.W. (ed.), *Structural and Chemical Organization of Teeth*, Academic Press, New York, pp. 45–110.

ØRVIG T. (1973), "Fossila fisktänder i svepelektronmikroskopet: gamla frågeställningar i ny belysning", *Fauna, Flora, Stockholm*, vol. 68, pp. 166–173.

PEYER B. (1968), *Comparative Odontology*, University of Chicago Press.

POOLE D.F.G. (1967), "Phylogeny of tooth tissue: enameloid and enamel in recent vertebrates, with a note of the history of cementum", in MILES A.E.W. (ed.), *Structural and Chemical Organization of Teeth*, Academic Press, New York, pp. 111–149.

POOLE D.F.G. (1971), "An introduction to the phylogeny of calcified tissue", in DAHLBERG A. (ed.), *Dental Morphology and Evolution*, University of Chicago Press, Chicago, IL, pp. 65–79.

PREUSCHOFT H., REIF W.E., MÜLLER W.H. (1974), "Funktionsanpassungen in Form und Struktur an Haifischzähnen", *Z Anat Entwickl-Gesch*, vol. 143, pp. 315–344.

PROSTAK K., SKOBE Z. (1986), "Ultrastructure of the dental epithelium and odontoblasts during enameloid matrix deposition in cichlid teeth", *Journal of Morphology*, vol. 187, pp. 169–172.

PROSTAK K., SKOBE Z. (1988), "Ultrastructure of odontogenetic cells during enameloid matrix synthesis in tooth buds from an elasmobranch, Raja erinacae", *The American Journal of Anatomy*, vol. 182, pp. 59–72.

PROSTAK K., SEIFERT P., SKOBE Z. (1990), "The effects of colchicine on the ultrastructure of odontogenetic cells in the common skate, Raja erinacae", *The American Journal of Anatomy*, vol. 189, pp. 77–91.

PROSTAK K., SEIFERT P., SKOBE Z. (1993), "Enameloid formation in two tetraodontiform fish species with high and low fluoride contents in enameloid", *Archives of Oral Biology*, vol. 38, pp. 1031–1044.

QU Q., HAITINA T., ZHU M. *et al.* (2015), "New genomic and fossil data illuminate the origin of enamel", *Nature*, vol. 526, pp. 108–111.

RADINSKY L. (1971), "Tooth histology as a taxonomic criterion for cartilaginous fishes", *Journal of Morphology*, vol. 109, pp. 73–92.

REIF W.-E. (1973), "Morphologie und Ultrastuktur des Hai-"Schmelzes"", *Zoologica Scripta*, vol. 2, pp. 231–250.

REIF W.-E. (1977), "Tooth enameloid as a taxonomic criterion: 1. A new euselachian shark from the Rhaeic-Liassic boundary", *Neues Jahrbuch für Geologie und Paläontologie, Monatshefte*, vol. 1977, pp. 565–576.

REIF W.-E. (1978), "Tooth enameloid as a taxonomic criterion: 2. Is Dalatias barnstonensis a squalomorphic shark", *Neues Jahrbuch für Geologie und Paläontologie, Monatshefte*, vol. 1978, pp. 42–58.

REIF W.-E. (1979), "Structural convergences between enameloid of actinopterygian teeth and of shark teeth", *Scanning Electron Microscopy*, vol. 2, pp. 546–554.

REIF W.-E. (1980), "Tooth enameloid as a taxonomic criterion: 3. A new primitive shark family from the Lower Keuper", *Neues Jahrbuch für Geologie und Paläontologie, Abhandlungen*, vol. 160, pp. 61–72.

RÜCKLIN M., DONOGHUE P.C. (2015), "*Romundina* and the evolutionary origin of teeth", *Biology Letters*, vol. 11, p. 20150326.

RÜCKLIN M., DONOGHUE P.C. (2016), "Reply to 'placoderms and the evolutionary origin of teeth': Burrow *et al.*, (2016)", *Biology Letters*, vol. 12, no. 20160526, pp. 1–3.

SANSOM I.J., SMITH M.P., MURDOCK D. *et al.* (1997), "Astraspis – the anatomy and histology of an Ordovician fish", *Palaeontology*, vol. 40, no. 3, pp. 625–644.

SASAGAWA I. (1989), "The fine structure of initial mineralizatin during tooth development in the gummy shark, *Mustelus manazo*, Elasmobranchia", *Journal of Anatomy*, vol. 164, pp. 175–187.

SASAGAWA I. (1993), "Differences in the development of tooth enameloid between elasmobranchs and teleosts", in KOBAYASHI I., MUTVEI H., SAHNI A. (eds), *Structure, Formation and Evolution of Fossil Hard Tissues*, Tokai University Press, Tokyo, pp. 113–121.

SASAGAWA I. (2002), "Fine structural and cytochemical observations of dental epithelial cells during the enameloid formation stages in red stingrays *Dasyatis akajei*", *Journal of Morphology*, vol. 252, pp. 170–182.

SASAGAWA I., AKAI J. (1992), "The fine structure of the enameloid matrix and initial mineralization during tooth development in the sting rays, *Dasyatis akajei* and *Urolophus aurantiacus*", *Microscopy*, vol. 41, pp. 242–252.

SASAGAWA I., ISHIYAMA M., YOKOSUKA H. *et al.* (2013), "Teeth and ganoid scales in *Polypterus* and *Lepisosteus,* the basic actinopterygian fish: an approach to understand the origin of the tooth enamel", *Journal of Oral Biosciences*, vol. 55, pp. 76–84.

SATHELL P.G., ANDERTON X., RYU O.H. *et al.* (2002), "Conservation and variation in enamel protein distribution during vertebrate tooth development", *The Journal of Experimental Zoology*, vol. 294, pp. 91–106.

SCHMIDT W.J. (1958), "Faserung und durodentin", *Metaplasie bei fischzähnen. Anatomischer Anzeiger*, vol. 105, pp. 349–360.

SCHULTZE H.-P. (2016), "Scales, enamel, cosmine, ganoine, and early osteichthyans", *Comptes Rendus Palevol*, vol. 15, nos 1–2, pp. 83–102.

SHARPE P.T. (2001), "Neural crest and tooth morphogenesis", *Advances in Dental Research*, vol. 15, pp. 4–7.

SHELLIS R.P. (1978), "The role of the inner dental epithelium in the formation of the teeth in fish", in BUTLER P.M., JOYSEY K.A. (eds), *Development, Function and Evolution of Teeth*, Academic Press, London, pp. 31–42.

SHELLIS R.P., BERKOVITZ B.K.B. (1976), "Observations on the dental anatomy of piranhas (Characidae) with special reference to tooth structure", *Journal of Zoology, London*, vol. 180, pp. 69–84.

SHELLIS R.P., MILES A.E.W. (1974), "Autoradiographic formatin of enameloid and dentine matrices in teleost fishes using tritiated amino acids", *Proceedings of the Royal Society*, vol. 53, pp. 51–72.

SHIMADA K., EVERHART M.J., DECKER R. *et al.* (2010), "A new skeletal remain of the durophagous shark, *Ptychodus mortoni*, from the Upper Cretaceous of North America: an indication of gigantic body size", *Cretaceous Research*, vol. 31, pp. 249–254.

SHIRAI S. (1996), "Phylogenetic interrelationships of neoselachians (Chondrichthyes, Euselachii)", in STIASSNY M.L.J., PARENTI L.R., JOHNSON G.D. (eds), *Interrelationships of Fishes*, Academic Press, San Diego, CA and London, pp. 9–34.

SIRE J.-Y. (1995), "Ganoine formation in the scales of primitive actinopterygian fishes, lepisosteids and polypterids", *Connective Tissue Research*, vol. 33, nos 1–3, pp. 213–222.

SIRE J.-Y., DAIT-BÉAL T., DELGADO S. *et al.* (2007), "The origin and evolution of enamel mineralization genes", *Cells, Tissues & Organs*, vol. 186, pp. 25–48.

SIRE J.-Y., DELGADO S., FROMENTIN D. *et al.* (2005), "Amelogenin: lessons from evolution", *Archives of Oral Biology*, vol. 50, pp. 205–212, 2005.

SIRE J.-Y., DONOGHUE P.C., VICKARYOUS M.K. (2009), "Origin and evolution of the integumentary skeleton in non-tetrapod vertebrates", *Journal of Anatomy*, vol. 214, pp. 409–440.

SIRE J.-Y., GÉRAUDIE J., MEUNIER F.J. *et al.* (1987), "On the origin of ganoine: histological and ultrastructural data on the experimental regeneration of the scales of *Calamoichthys calabaricus* (Osteichthyes, Brachyopterygii, Polypteridae)", *The American Journal of Anatomy*, vol. 180, pp. 391–402.

SMITH M.M., HALL B.K. (1990), "Development and evolutionary origins of vertebrate skeletogenic and odontogenic tissues", *Biological Reviews of the Cambridge Philosophical Society*, vol. 65, no. 3, pp. 277–373.

STEWART J.D. (1980), "Reevaluation of the phylogenetic position of the Ptychodontidae", *Transactions of the Kansas Academy of Science*, vol. 83, no. 3, p. 154.

SUMMERS A.P. (2000), "Stiffening the stingray skeleton – an investigation of durophagy in myliobatoid stingrays (Chondrichthyes, Batoidea, Myliobatidae)", *Journal of Morphology*, vol. 243, pp. 113–126.

THIES D. (1982), "A neoselachian shark tooth from the Lower Triassic of the Kocaeli (= Bithynian) Peninsula, W Turkey", *Neues Jahrbuch für Geologie und Paläontologie, Monatshefte*, vol. 5, pp. 272–278.

THOMASSET J.J. (1930), "Recherches sur les tissus dentaires des poissons fossiles", *Archives d'Anatomie. Histologie et Embryologie*, vol. 11, pp. 5–153.

TOMES C.S. (1898), "Upon the structure and development of the enamel in elasmobranch fishes", *Philosophical Transactions of the Royal Society*, vol. 190, pp. 443–464.

TURNER S., BURROW C.J. (2011), "A Lower Carboniferous xenacanthiform shark from Australia", *Journal of Vertebrate Paleontology*, vol. 31, no. 2, pp. 241–257.

UNDERWOOD C.J. (2006), "Diversification of the Neoselachii (Chondrichthyes) during the Jurassic", *Paleobiology*, vol. 32, pp. 215–235.

VENKATESH B., LEE A.P., RAVI V. *et al.* (2014), "Elephant shark genome provides unique insights into gnathostome evolution", *Nature*, vol. 505, pp. 174–179.

VÉLEZ-ZUAZO X., AGNARSSON I. (2011), "Shark tales: a molecular species-level phylogeny of sharks (Selachimorpha, Chondrichthyes)", *Molecular Phylogenetics and Evolution*, vol. 58, pp. 207–217.

WELTON B.J., ZINSMEISTER W.J. (1980), "Eocene neoselachians from the La Meseta Formation, Seymour Island, Antarctic Peninsula", *Contributions in Science, Natural History Museum of Los Angeles County*, vol. 329, pp. 1–10.

WITTEN P.E., SIRE J.-Y., HUYSSEUNE A. (2014), "Old, new and new–old concepts about the evolution of teeth", *Journal of Applied Ichthyology*, vol. 30, pp. 636–642.

WOODWARD A.S. (1889), *Catalogue of the Fossil Fishes in the British Museum (Natural History). Part 1*, British Museum (Natural History), London.

ZHU M., YU X., AHLBERG P.E. *et al.* (2013), "A Silurian placoderm with osteichthyan-like marginal jaw bones", *Nature*, vol. 502, pp. 188–193, 2013.

ZANGERL R. (1981), *Handbook of Paleoichthyology Volume 3A-Chondrichthyes I Paleozoic Elasmobranchii*, Verlag Dr. Friedrich Pfeil.

ZANGERL R., WINER H.F., HANSEN M.C. (1993), "Comparative microscopic dental anatomy in the Petalodontida (Chondrichthyes, Elasmobranchii)", *Fieldiana Geology*, vol. 26, p. 142.

ZYLBERBERG L., SIRE J.-Y., NANCI A. (1997), "Immunodetection of amelogenin-like proteins in the ganoine of experimentally regenerating scales of *Calamoichthys calabaricus*, a primitive actinopterygian fish", *The Anatomical Record*, vol. 249, pp. 86–95.

Index

A, B, C

acrodin, 16, 18, 81, 83–85, 88–91, 93–95
actinopterygians, 16, 17, 81, 83, 85, 88–91
ameloblasts, 2, 8–14, 20, 25, 27, 64, 83, 92
anachronistidae, 69
ancestral, 20, 26, 28, 29, 34, 35, 53, 55, 69
batomorphs, 3, 6, 7, 33, 47, 51, 53, 54, 56, 57, 59, 61, 62, 69, 72, 79
biomechanical, 92
chondrichthyes, 2–4, 14–16, 18, 19, 23, 24, 27
convergently, 90, 95
cutting tooth, 69, 75

D, E, F

denticle, 11, 28, 38
dentine, 2, 13, 14, 17, 38, 41, 42, 44, 45, 57, 78, 79, 93

dentition, 1, 10, 27, 40, 41, 54, 55, 57, 62, 71, 73, 76, 77, 85, 90
development, 7, 9, 11–13, 39, 41, 76, 83
durophagous, 4, 42, 48, 51, 54, 55, 57, 58, 61, 78, 79
elasmobranchs, 5, 6, 33, 81, 83, 84, 88, 91
electric rays, 53
fossil, 1, 3, 19, 26, 36, 44, 48, 51, 54–56, 69, 71, 73, 74, 76, 79, 83, 84, 89, 91, 92

G, H, J

ganoine, 81, 83
grinding, 14, 41–45, 70, 77, 78, 85, 89, 90, 95
Heterodontus, 2, 4, 42, 44, 58, 77, 78, 85
histology, 3, 10, 14, 48, 51, 53, 59
homology, 7, 16, 55, 81, 84, 91

hybodonts, 5, 6, 33–37, 39–
    41, 43, 45, 53, 71–73, 76,
    78, 88, 90
jaws, 26, 62, 71, 83, 90, 93

M, N, O

microphagous, 61
mineralization, 2, 7–10, 14,
    16, 17, 24, 65, 68, 92
Mobulidae, 54–57
morphology, 13, 14, 19–21,
    24, 29, 33, 34, 43–45, 48,
    55–57, 62, 69, 71, 73, 76–
    78, 86
Myliobatiformes, 13, 52, 54–
    59, 79
neoselachians, 3, 5, 6, 15, 16,
    19, 33–35, 37, 47, 53, 55,
    68, 71
odontoblasts, 2, 8, 9, 11–13,
    16, 20, 65, 83, 89, 90, 92
ostracoderms, 16, 27

P, R, S, T

Palaeozoic, 73, 76, 95
phoebodonts, 23–27
phylogenies, 47
placoderms, 15, 92–95
planktivorous, 54, 55, 58
predation, 61, 66, 73, 76, 79
primitive, 20, 21, 24, 26, 31,
    34, 35
pycnodonts, 89, 90
ray, 43, 51, 53, 54, 81, 92
selachimorph, 10, 21, 30, 34,
    42, 59, 61, 69, 76, 77, 79,
    86, 90
shark, 19, 28, 35, 40, 43, 61,
    62, 67, 74, 77, 78, 83, 90
squalomorph, 68
stresses, 39, 40, 42, 54, 56–
    58, 62, 77–79, 88, 90, 92
teleosts, 92

Printed in the United States
By Bookmasters